*"On Second Thought" and
Other Essays in the History
of Medicine and Science*

"On Second Thought" and Other Essays in the History of Medicine and Science

⤙⤙ OWSEI TEMKIN ⤚⤚

The Johns Hopkins University Press
Baltimore and London

© 2002 The Johns Hopkins University Press
All rights reserved. Published 2002
Printed in the United States of America on acid-free paper
2 4 6 8 9 7 5 3 1

The Johns Hopkins University Press
2715 North Charles Street
Baltimore, Maryland 21218-4363
www.press.jhu.edu

Library of Congress Cataloging-in-Publication Data

Temkin, Owsei, 1902–
"On second thought" and other essays in the history of medicine and science /
Owsei Temkin.
p.   cm.
Includes bibliographical references and index.
ISBN 0-8018-6774-6 (alk. paper)
1. Medicine—History.   2. Science—History.   I. Title.
R131.T42   2002
610'.9—DC21   2001029420

A catalog record for this book is available from the British Library.

To the memory of my wife,
Clarice Lilian Temkin, née Shelley, 1906–1992

Contents

Contents

Preface and Acknowledgments

More than twenty years have passed since I published a selection of my essays under the title *The Double Face of Janus.* Insofar as the present publication offers an additional selection, it is a sequel to *The Double Face.* Some of these essays were excluded because of a limitation of space in the previous work; others were written later, and of these some are second thoughts deviating from earlier beliefs. This is particularly true of the interpretation (and translation) of the Hippocratic Oath, an interpretation that forced itself upon my mind in recent years and led to my writing the first two essays in this collection: "What Does the Hippocratic Oath Say?" (Chap. 2) and "On Second Thought," the introductory essay whose title serves also for the whole volume. The reprinted articles have undergone only minor stylistic changes to give the book a uniform character.

In *The Double Face of Janus,* I tried to coordinate my work with events of my life. The events that have taken place since then deserve no more than a gerontological summary. Forgetfulness, legs that restrict me to a wheelchair, and fingers that refuse to do their job, above all writing properly: all these have conspired to isolate me from libraries, including my own, and prevent me from consulting the literature, citing exact references, and giving credit to all those to whom credit is due. In short, they have put an end to the kind of work I used to do. I must therefore ask for the goodwill of the readers of the introductory essay and of the chapter on the Hippocratic Oath.

↠ • ↞

Were it not for the editorial help of my daughter, Judith Temkin Irvine, this book would never have seen the light of day. I am indebted to the Johns Hopkins University Institute of the History of Medicine and its

director, Dr. Gert Brieger, for a grant for the preparation of an index and for help in obtaining permissions for reprinting the articles. I am grateful also for the help and patience of the Johns Hopkins University Press, especially Executive Editor Jacqueline Wehmueller and Senior Manuscript Editor Linda Forlifer.

*"On Second Thought" and
Other Essays in the History
of Medicine and Science*

CHAPTER I

On Second Thought

Old Age and the Historian

Gaudeamus igitur iuvenes dum sumus, "so let us enjoy ourselves while we are still young." And the famous student song continues: "After pleasant youth, after miserable old age, earth will have us." The song is right, old age carries its troubles. The late Erwin Ackerknecht used to refer to the many frailties of advanced age as "the penalty for getting old." But old age is not without its compensations. The freedom from the hustle and bustle of life and the reduced nightly sleep allow more time for thinking, and if one's reflections turn to one's own past activities, they may lead to second thoughts. I even venture to recommend old age to the historian, who can now reflect on the time of which he was an active witness and about which younger historians speak merely on the basis of sources. Such an experience can be instructive.

I lived in Germany through almost the whole time of the democratic Weimar Republic. When I now hear younger people's comments on "the Weimar culture," I often learn things of which I had not been aware, but I sometimes have the impression of dealing with people who speak learnedly about roses without ever having smelled one. When it comes to earlier periods, we all are in the position of such scholars, and a few telling anecdotes can do more for our empathy, our feeling for the period, than can a closely argued thesis.

Investigating Ancient Medicine

Two men, both of them deceased by now, have had a decisive influence on my life as a scholar: my teacher and chief, Henry E. Sigerist, and my friend and colleague, Ludwig Edelstein. The former not only launched me upon

my academic career and made me see medicine as part of the cultural and social life of a period but also, by taking me along from Germany to the United States in 1932, may have saved my life. From Edelstein I learned much about the art of interpreting a text. For many years our offices at the Johns Hopkins Institute of the History of Medicine were located door by door, and more often than not, the doors were open, thus facilitating the exchange of ideas and news of the day. When Edelstein wished to curtail his heavy smoking, we arranged that he would keep his cigarettes in a drawer of my desk so that he would have to come to me every time he wished to light up. (The experiment was not successful. Eventually he removed the obstacle he had put in his way and kept his cigarettes to himself.) Since I was no longer smoking cigarettes, I could not reciprocate, but I could keep abreast of his work and of his thinking, benefit from it, and occasionally contribute some comments. For many years Edelstein's views of ancient medicine became also largely my own, though I could not accept his dogmatic conviction that historical research would be pointless were it not to aim at absolute truth. I have been satisfied with probability, especially in the field of antiquity, where only a fraction of the sources has survived.

I shall come back to my relationship to Sigerist, but first I will say something about my ever increasing distance from Edelstein's position. I have always considered myself a medical historian. Edelstein's interest lay in philosophy and Greek culture, and he approached medicine as part of those two. This approach shows in most of his work. The ancient physician was not the colleague of the modern one. He received his scientific notions from philosophy (which included "science"), elaborating them and adapting them to the needs of his art. What was true of medical theories, according to Edelstein, was equally true of medical ethics, as most clearly shown in the Hippocratic Oath, which Edelstein interpreted as a "Pythagorean document."

Edelstein's emphasis on the "otherness" of ancient medicine was very wholesome at the time, even if it was exaggerated and thus likely to distract attention from specifically medical matters. This became increasingly clear to me while I was working on my book, *Hippocrates in a World of Pagans and Christians*. Edelstein had emphasized Stoic influence on Scribonius Largus (fl. around A.D. 30), a minor Roman physician and author of a little pharmacological text, and on the Hippocratic *Precepts* (first century A.D.?).[1]

In both cases, this Stoic influence is very likely. Doctors, too, are children of their time, and the Stoic philosophy was very influential in the educated stratum of society during the first century. But Stoicism does not explain why a little prescription book should be prefaced by a lengthy dedicatory chapter on medical ethics. The inclusion of such a preface is understandable, however, if we realize that the booklet was intended for the laity and that the preface was "to dispel Roman fears of poisoning and of physicians' experimenting on the patients."[2]

As to *Precepts*, an analysis of Stoic "philanthropy" does not lead far beyond generalities. But a consideration of the ethical behavior that *Precepts* teaches shows it not very different from the attitude a modern doctor is expected to display toward patients and colleagues.[3] Obviously, in speaking of the medicine of cultures different from ours, we have to reckon with its otherness as well as with those elements that subsume it under the common denominator of "medicine." It seems to me that it is by comparison with modern practices that a clearer picture will emerge, rather than by appealing to abstractions and generalizations.[4]

My distancing from Edelstein has reached its acme in my translation and interpretation of the Hippocratic Oath, to which the second chapter of this book is devoted. I valued Edelstein's work, even when support for his thesis of the oath as a Pythagorean document was weakening, because his view gave meaning to the document as a whole—which the criticism, even where justified in detail, often did not. But finally it became clear to me that the thesis stood or fell with the historical existence of a group of Pythagorean physicians, a group for which independent testimony was very weak; on the other hand, the text of the oath makes sense without any auxiliary hypotheses, as my essay tries to show. Of course, every translation implies an interpretation, that is, its understanding by the translator. Thus, Edelstein, with a translation that makes the juror swear to guard his life and his art "in purity and in *holiness*" (my emphasis), was already on the way to Pythagoreanism, whereas my translation, "in purity and in *piety*," facilitates a less esoteric understanding.

However, Edelstein and I both face the question: what do you mean by claiming that the Greek text that you translate is the oath ascribed to Hippocrates? We both use a modern edition constructed on the basis of available manuscripts, which are copies of older manuscripts, so that at best the edition represents a text at the basis of all the manuscripts known

today. With this text we assume that we have come as close as is possible to what we call the Hippocratic Oath. Every edition has offered a more or less different text, and every new edition will also do so, possibly with the help of new material (papyri? translation? commentaries?) that will give us a better idea of how to fathom the origin of the oath and the form in which it appeared as a written text.

In spite of the uncertainty besetting the oath, there is good reason to ask what it says. The oath incorporates the idea that a doctor should be bound to a moral code. In the modern debates pro and con swearing such an oath—pro and con accepting the Hippocratic Oath as we know it, keeping it as is, modifying it, or abandoning it altogether—we should at least know what we are talking about. And *we* includes the public when it appeals to the Oath of Hippocrates.

I think Edelstein would have agreed with that, and it is only fair to concede to him that Pythagorean influences might well have played a role *behind* the oath, even if it is not read as a Pythagorean document as such.

To atone for my heresy, I have included in this volume the obituary of my friend Edelstein, a statement written before I developed second thoughts about his approach to ancient medicine.

There are some other second thoughts about research on ancient medicine that have lately come to me. If, to clarify the attitudes of doctors in former periods, it is desirable to compare them with the attitudes prevailing today, we must know *our* expectations and demands vis-à-vis patients and society. I am not thinking of the knowledge the doctor should possess; some kind of knowledge always was expected, though it changed from age to age. Rather, I am thinking of behavior and feeling. This is not the simple matter it has seemed to be.

The ethical code of the Hippocratic Oath has been a paradigm for the doctor's duty and responsibility. They can be taught, learned, codified, and legalized. As is probably true of all human actions, acting responsibly may also have an emotional undertone: the doctor who acts irresponsibly may be pricked by his conscience.

But where the doctor's feeling for his patient is concerned, the example usually cited is the parable of the good Samaritan who cared for a wounded stranger because he loved his neighbor like himself. In the Greco-Roman world, too, certainly by the first century A.D., compassion was ex-

pected from the doctor and from a good midwife.[5] In "The Idea of Respect for Life in the History of Medicine" (Chap. 3), I sketched several historical situations up into modern times where the idea of respect for life has manifested its various facets. This chapter touches on the doctor's duty and compassion, which, though different things, are not always easily separated. Their difference emerges most clearly when they conflict. Thus, the Roman encyclopedist Celsus (first century A.D.) thought "that a surgeon should be 'filled with pity so that he wishes to cure his patient' without, however, being moved by the patient's cries to too great a hurry or to cutting less than required."[6]

Some eighteen hundred years later, Osler, in a similar vein, told medical students to "cultivate, gentlemen, such a judicious measure of obtuseness as will enable you to meet the exigencies of practice with firmness and courage, without, at the same time, hardening 'the human heart by which we live.'"[7] That advice meant that, in the practice of medicine, compassion tempered by imperturbability and equanimity should motivate the doctor to do what has to be done and perhaps even more.[8]

I quote Osler because, in the Anglo-American world at least, his teaching and example have been accepted as a desirable model for the doctor. However, the question is: are they still being lived up to?

Osler also wrote: "The practice of medicine is not a business and can never be one." With all due respect to Osler, medicine in his day also had its business side, even if it might be covered by slipping an honorarium as a token of gratitude into the hand of the doctor who would never bill his patients. Arguments were loud in antiquity as to whether gain or philanthropy ought to rule the doctor. There also exists the old Latin saying: *dat Galenus opes,* "Galen (i.e., medicine) gives riches," and riches are not acquired by solely doing one's duty and being compassionate. But business as a sideline is different from medicine as a business, where medical care is dispensed "as man deals in corn and coal," to quote Osler again.[9] Along with many contemporary commentators, I wonder whether we, in the United States, are not coming closer to medicine's becoming a business in that sense? Health insurance organizations seem to limit the doctor's responsibility to his patients, the rush of malpractice suits provokes unnecessary diagnostic and therapeutic actions, and compassion is in danger of becoming a luxury that not everybody can afford. Such a

trend is likely to be strengthened by economic considerations that force medical schools with an emphasis on research to replace the openness of science and therapy with a search for patents and efforts at marketing.

These things may change, and I do not mention them to criticize or to complain that "the country is going to the dogs"—a complaint that is the hallowed privilege of old people. For better or for worse, if this trend continues it may end by producing doctors who think of themselves as persons engaged in a business and who feel and act accordingly. Rather than criticizing, I want to make the point that, in comparing the medicine of past periods with medicine today, it is not enough to consider the undoubted progress that has been made in medicine's scientific, technical, and social aspects. It is also necessary to try to fathom how doctors, nurses, medical researchers, and administrators thought and felt about their relationship to the sick. I say "try to fathom" because, in the final analysis, nobody can look into another's heart, probably not even into his own.

Historiography and Biography: The Individual and Its Potencies

Sigerist was my teacher, and his influence probably was more pervasive than I am aware of. It was part of his liberality and his tolerance for other people's individuality and freedom that, as a chief, he allowed his co-workers to go their own ways and to disagree with him, even when his attention had turned to medical sociology and political issues connected with it. In two directions my professional debt to Sigerist is obvious to me: he made me aware of medicine as part of our cultural and social life, and in his lecture course in the history of medicine, which he offered at the University of Leipzig, he gave me a view of that history as a development from prehistory, Egypt and Mesopotamia, through Greco-Roman antiquity, Arabic and Latin Middle Ages, Renaissance and the other historical periods of the European West down to the dawn of the twentieth century. Here I learned the names of the great doctors, their contributions, and the medical ideas and tendencies of the periods (e.g., therapeutic nihilism; the anatomical idea). This course showed much originality;[10] nevertheless, it is fair to characterize it as of the type represented by the leading textbooks of the twenties.

Lingering doubts about a strict coordination between culture and med-

ical thinking were strengthened when I found that the quantitative reasoning in William Harvey's *De Motu Cordis* had a predecessor and probable model in one of Galen's major works.[11] But a serious revision of my concept of history and historiography started in 1973 while I was working on "Science and Society in the Age of Copernicus" (Chap. 5), a paper that was occasioned by the celebration of the fifth centenary of the birth of Nicolas Copernicus (1473–1543). The seventy years of his life crossed many of the conventionally recognized historical periods: Renaissance, Reformation, Humanism, the European discovery of America and of the sea route to India and to the Far East, and others. Yet his life span belonged to none of these periods exclusively. Many scientific (in the modern sense) data can be assigned to the age of Copernicus, yet they were not neatly dissociated from religious, magical, aesthetic, and moral ideas. In short, this section out of the human past demonstrated "the untidiness of human affairs," an expression I had borrowed from an author whose name has slipped my mind.

I do not remember the steps that have led me to think of "the past" as amorphous, devoid of any structure. Structure is given to it by us. Motivated by our interests and our religious and metaphysical convictions, we use historiographical forms such as chronicles, annals, archaeological strata, cycles, progress, chronology, evolution, revolution, "great men (or women)," and so on. Thus history is created or, I should say, histories, for "the" history, as something absolute, may exist for God—thought of as absolute, timeless truth—but not for the human race. The chronologies that serve as a kind of set of historical coordinates are due to the religious motive of choosing as their zero point an event believed crucial for the respective religion: the creation of the world, the birth of Christ, or Mohammed's flight from Mecca to Medina.

"We" in this connection may range from the individual to entire groups and populations, and "interests" may be intellectual, esthetic, material, religious, altruistic, power-hungering, or anything else that drives human beings. There is nothing new in thus connecting the exploration of the past with our living interests. I emphasize it because of its bearing on what I have to say on the study and the teaching of medical history.

It may be rightly questioned how the essay on science and society in the age of Copernicus fits into what I have just said. It fits badly, because its format is that of a biography rather than that of a history. The seventy

years of Copernicus's life are its time, and scientific events take the place of the data in a biographee's life. Historiography and biography are so closely related that biographers can hardly avoid referring to the historical background of their subject's life, while historians—unless they are radical sociologists—cannot do without entering upon the lives of the leading figures of their tale.

Medical historians also write biographies, usually of eminent doctors, or they write pathographies, usually of eminent persons, such as Churchill, Chopin, or van Gogh, whose illnesses the author discusses. But practicing physicians are also biographers of a sort, when in writing case histories they set down their patients' lives from the point of view of health and disease. The psychiatric case history makes this especially clear.

The medical case history, like biography in general, is the form in which an individual can be captured conceptually, in contrast to science, which deals with the statistically available (or, in the prestatistical age, with a plurality of cases). Thus, science deals with diseases and sick people (in the plural), while the case history deals with the individual patient. The illness of George (or Nancy) becomes a "case" of the disease (or syndrome) X. Even a molecule needs the history of its identification to become recognizable as one particular molecule.

Individuum est ineffabile: The individual is ineffable; it cannot be made to fit into categories. There is no science of the individual. Fingerprints belong to science, but everybody has his or her individual fingerprints. Moreover, human individuals have introspection: I see red, but I do not know whether anybody else has exactly the same visual experience. If, nevertheless, there is a science of optics, it rests on the fact that what I call *red* has the same physical, chemical, biological, and social (e.g., as "red" on traffic lights has a conventional meaning) attributes for everybody who claims to see red. But all these attributes, which even a blind person may understand, will not make that person see red.

Thus the human individual is a composite of two potencies. Subjectively he or she sees, feels, hears, smells, and tastes, wishes, desires, chooses, decides, and is emotionally moved. At the same time he or she has the ability to think objectively, being endowed with reason. In the living individual the two potencies are mutually dependent. When I think of myself, I refer introspectively to my ego, but at the same time I am making an object of the introspective ego. If the neuroscientist could not see and

feel the brain, there would be no neuroscience, but, obviously, there would also be no neuroscience if neuroscientists could not detach themselves from the subjective basis of their work. When the individual becomes "a neuroscientist," his or her work can be judged publicly. In the process of detachment, however, scientists are liable to forget that researchers are individuals and that the subjective, introspective potency remains a partner of their work.

Perhaps the acme of this detachment is reached when an author, referring to himself or herself, speaks of "the author of the present work." But the introspective, subjective partner may, in turn, insist on its rights, possibly even aggressively so. The "mind-body problem," and the uneasy co-existence of doctor and psychotherapist, illustrate the old difficulty of combining what is subjective and objective, or individual and public, in our lives. "Gall and the Phrenological Movement" (Chap. 6 in this volume) deals with an early, mid-nineteenth century example of the conflict between the objective and the subjective sides of human beings. By its followers phrenology as the "science of mind" was welcomed as replacing introspection by objective science. Its enemies, on the other hand, called it materialistic and hostile to religion and morality.

I have lately come to look at another facet of the age-old conflict. As Celsus (first century A.D.) reports, the Dogmatists, a medical sect that favored anatomical dissection and physiological study, praised Herophilus and Erasistratus (famous members of the sect) for having cut open living criminals whom the king had turned over to them from the jail.[12] They defended this act of human vivisection against the accusation of cruelty by alleging that it "was not cruel if the suffering of criminals—and moreover only a few of them—would serve the search for remedies for innocent people throughout the ages."[13] It is a moot question whether these vivisections really took place; the denial probably is largely due to our reluctance to credit doctors with such cruelty.[14] The Dogmatists, too, did not deny that by itself vivisection was cruel, though they tried to blunt compassion by emphasizing that the victims had been criminals. But their argument against the imputation of cruelty was that the advantage for innocent people of all times outweighed the suffering of a few criminals. Innocence deserves sympathy, an emotional factor, and the appeal to the benefit of the many cannot blot out the cruelty to the few; it can only hope to make the public disregard it. I am here reminded of what I some-

where read as attributed to Stalin: "The death of one individual is a tragedy, the death of a thousand people is a statistic."

Human vivisection is not an issue any longer, but the conflict between individual and public has, if anything, even broadened. We have only to think of the conflicts between the interests of the individual patient and the demands of public health. In the context of second thoughts, I remember the forceful insistence of a colleague that the large cost of dialysis, which benefited only relatively few people, would better be used for medical research. It has occurred to me only now that I ought to have asked my colleague whether he would still defend that opinion if his mother were one of those from whom dialysis was to be withheld. His answer might have given me a hint in which direction the wind was blowing: compassion for the individual or cost-effectiveness and public advantage.

When I was young, medicine was dominated by the doctor, helped by the nurses, who treated and cared for sick people. Today we speak of "the health field" and "the health industry," of which medicine, in the sense in which I have been using the term, is but a sector and the doctor is but one of many "health professionals." For the "health industry" it seems natural to think of medicine as a business and of the doctor as a "provider." Compassion for the individual does not enter into the ledger. On the other hand, the industry may claim that it helps to make available new and better medicines and medical care for thousands who formerly could not afford it.

I should like to conclude this section with some general remarks on my use of the terms *individuality* and *individualism.* They need clarification in two directions. *Individuality,* as I used the term, has nothing to do with the egoistical individualism that is antisocial, looking for personal interest only. I trust that my comments on compassion will have made that clear. On the other hand, a social conscience that looks for feeding and housing for all who are in need and seeks to provide them with medical care may spring from the same motive that made the good Samaritan help the wounded stranger, but it may also be the outgrowth of ideas of fair play and social justice—ideas that belong to realms other than medicine.

On Teaching Medical History

In 1929, in an address delivered at the dedication of the Institute of the History of Medicine at the Johns Hopkins University School of Medicine, Harvey Cushing raised the following question: "Will this foundation merely mean still another group of specialists having their own societies, organs of publication, separate places of meeting, separate congresses, national and international, and who will also incline to hold aloof from the army of doctors made and in the making?"[15]

This question was meant as a warning that medical history might become another medical specialty out of touch with the mainstream of the medical profession. What was meant as a warning proved to be a prophecy. All the points mentioned by Cushing have become realized, and more than that: medical history in the United States is on the verge of becoming a historical specialty. This trend has gone so far that a few years ago I was surprised to find a Ph.D. affixed to my name on the directory of the Johns Hopkins Institute. It was obviously taken for granted that, being a medical historian, I must have that degree. And I am somewhat saddened to realize that, judging by the requirements for academic positions at some of our foremost departments of medical history, I would not even make it to an assistant professorship, my medical degree counting for nothing, since not followed by a Ph.D.

I hasten to emphasize that I have no intention of weighing the value of degrees for the academic history of medicine. Competence in historical research is indispensable. But how it has been acquired is irrelevant and should not be determined by academic formality. Rather, I have mentioned degrees because they symbolize the shift that has taken place not only from medicine to "the health field," on which I have already remarked, but also from a limited and passive interest in medical matters on the part of the public at large to an active interest that manifests itself in discussions about health insurance, in the frequency of malpractice suits, subscription to health letters, jogging, attention to dietary matters, and so on. The efficacy of medicine has wonderfully increased, workers in the health field are proud of it, people live healthier and longer lives, they *expect*—if not *demand*—their doctors to treat them effectively, they read of new "breakthroughs," which their newspapers report almost daily, and they are anxious to be insured, for the maintenance of good health has become

very costly for both patients and providers (doctors, hospitals, medical schools, laboratories, pharmaceutical firms).

Public ignorance of medical matters certainly is not desirable, and achievements in preventive medicine are highly welcome. But I cannot altogether banish the impression that we may be going the same way as ancient dietetic medicine went when the fear of disturbing the harmonious balance of the body's humors and qualities put healthy people under steady medical surveillance, and health became the *summum bonum* for which they lived.[16] To quote myself: "Most of us live in a state of compromise. We believe in values beyond health, yet we also sometimes forget them and respect our lives more than anything else." On second thought, I now add that reading Molière's *Imaginary Patient* might help to prevent undue preoccupation with health over proper regard for it.

Medicine has become a subject of broad public interest. Something that is of such wide concern invites serious historical study, which is no longer to be left in the hands of doctors. Thus, the history of medicine has become a historical discipline in line with others, such as the history of art, of music, and of law. All these disciplines demand dedication to the craft of historiography, while the relationship to the particular subject is left open. Edelstein was an outstanding historian of medicine with little interest in actual medicine and with even less medical knowledge.

"The Study of the History of Medicine" (Chap. 14 in this volume) marks my position when the relation of history to medicine had entered a problematic state. But as the controversy over medical history between the clinician Wunderlich and the medical historian Haeser—a controversy that took place about a hundred fifty years ago—shows (see Chap. 15), the issue is an old one. Fundamentally it is an issue faced by all basic medical sciences in relation to their mother sciences (e.g., medical physiology to physiology), but it is sharpened by the fact that history is not a natural science.

George Rosen spoke of iatrocentric historiography, which placed doctors and the medicine they taught and practiced in the center of the history of medicine. Most history of medicine down to the 1920s and 1930s was iatrocentric, including the kind offered by Sigerist and the textbooks. Nothing could be more iatrocentric than the Oslerian method of teaching medical history: Let the professor of medicine in his clinical instruction acquaint his students with the lives of great doctors and the classi-

cal descriptions of diseases. From a general historical point of view, "ia-trocentric" represents a limitation.

As a branch of history, history of medicine will avoid iatrocentric one-sidedness. It will embrace all that belongs to the health field and it will consider the cultural, economic, environmental, and political conditions that impinged on the medicine of a period and of a country. I think that Sigerist's *History of Medicine*, judging by the two volumes that appeared, tended to this kind of broadly conceived history of medicine.

History of medicine will do justice to the individual character of the periods that historiography customarily recognizes and, if chronologically arranged, it will lead to the history of medicine of our time—that is, medicine as it is taught, practiced, and publicly accepted in the historian's time and country—and considered "modern medicine." This kind of history easily leads to a comparative treatment, the usefulness of which I indicated when I suggested that attitudes and actions of doctors of past times be compared with those of modern doctors. But this needs a caveat: comparison must not slip into the relativistic fallacy of taking modern medicine for "just another form" of medicine on a par with past and future forms.

I shall try to explain my meaning by referring to an episode in Dr. Sigerist's seminar on "The Wonder."[17] A medical student (who happened to be the son of the professor of forensic medicine at the University of Leipzig) read a paper on the miracle of Francis of Assisi, in which he tried to explain the saint's stigmatization (i.e., the appearance of bleeding spots on his hands and feet) physiologically. A professor from the philosophical faculty who also attended the seminar made short shrift of the paper. Our sources for this stigmatization (he pointed out) knew nothing of red blood corpuscles, and so on. To them the wonder was the *Imitatio Christi*, the appearance of the bloody wounds of the crucified Christ. The professor was right in accepting the historical interpretation of the stigmata of Francis. But he was wrong in brushing the student's paper aside as an anachronistic blunder. Being a man of the twentieth century, the professor entrusted his health and his life to the application of modern science, and modern science has to explain cases of stigmatization as best it can; it cannot accept miracles without renouncing the invariance of natural laws.

Medicine and science can be studied as bound to the culture of their particular time, and the same is true of our own day. Thus, it would seem that all these activities are equally relative. But this means overlooking a

fundamental fact. Just as science depends on scientists, all of whom are individuals, so medicine depends on individuals who by their lives, and by the values by which they live, testify to what they deem true and valid. As rational beings we humans can and do view many things relatively, and we may be influenced thereby. But being individuals we cannot and do not live in a relativistic world. We cannot deny our sensations, our sentiments, our urges, and all those things whose acceptance we prove by our actions, just as we cannot and do not live without judging and believing some things to be true and others to be false.

Thus far I have spoken of history of medicine as a historical discipline. Harvey Cushing, however, thought of it as primarily a medical affair, as did most other people, lay as well as medical, Dr. Sigerist and myself included, in 1929 and in 1946 when I published my "Essay on the Usefulness of Medical History for Medicine." Still, as long as few nonmedical people engaged in medical history, there was little discussion as to which of the two—medicine or history—should determine the character of the discipline. As mentioned above, the development points in the direction of history, and that poses an important question: should medical history still have a place in schools of medicine? Should it be taught to medical students?

In the essay just mentioned, I tried to make a case for the usefulness of medical history for medicine. But that was more than fifty years ago, and teaching must be adjusted to the times. In our day students of medicine have to meet a crowded and demanding curriculum. There are so many things that must be taught that a harassed dean may well ask "What will seriously be missed if no medical history course is offered?" Answers will vary, and I can only state my personal opinion: "Just because modern medicine is proving very successful and promising, a humanistic counterweight to the claims of allegedly irresistible developments would be missing, a counterweight other than the one that clinical teaching, by its encounter with patients, provides."

"What will such a historical course offer?" asks the dean. I take a deep breath, look fearlessly into the dean's eyes and present my case.

"To avoid disorientation and confusion, the course must be built on a scaffold of the history of medicine such as I suggested. The students should then understand how we arrived at modern medicine, which the school expects them to know and later on, in one form or another, to prac-

tice. But what I had in mind when I mentioned a humanistic counter-weight are problems of aims and attitudes, which history brings to the fore and which ask for human reactions.

"As long as cesarian section practically assured the mother's death, moral and religious convictions determined whether mother or child should, if possible, be saved.[18] On the other hand, whether abortion is justifiable was and is primarily an ethical problem. To give other examples: What were the motives leading people to become doctors—to devote themselves to helping the sick, to pursue medical science, or to make money? What should have priority, the needs of the individual or public welfare? What led some doctors, like John Hunter and Jesse Lazare, to experiment on themselves at the expense of their own health? What inspired men like Semmelweiss and Freud to stand firm in spite of the resistance and ridicule that confronted their work? Thus, the course would not only show the student that many of the problems facing us today also faced past generations under conditions different from ours but also illustrate how differently doctors understood their task and shaped their lives and how they have become models for possible emulation or rejection."

Interrupting me at this point, the dean asks, "Do you mean to say that your historian should answer this kind of question and also tell the students which models to follow?"

"No. The historian's task is to tell what was and what has been and what has led to present conditions. As historian he or she is neither a prophet nor an ethicist, nor, generally speaking, a preacher of what should be done . . ."

"Well then, what good is all this to do? Where is the need for your course, and why should precious time be allotted to it?"

"Because the course is supposed to encourage the students' confidence that they can be masters of their own fate and of the fate of their profession. As a Latin saying has it: *Tempora mutantur, nos et mutamur in illis*— 'Times change and we change with them.' Even in the few short years of medical school, the students experience change. Modern medicine is not static; not everything that held true when they entered medical school universally holds true when they leave it. History expands that horizon over thousands of years; the student becomes the heir of knowledge and skill acquired over the centuries, but also the heir of the kinds of problems that are not matters of positive knowledge and skill. Instead, these kinds of

problems ask for reactions to particular events that are at issue. The reactions can vary, and that means that we all have some degree of freedom to decide our fate. This freedom has nothing to do with metaphysical problems of the freedom of will. It is the freedom that every individual exercises in the affairs of daily life, even if he or she does not believe in free will at all.[19]

"I think that such a course would be timely because I see medicine as challenged by such trends as I mentioned before and which demand some kind of response, be it assent, resistance, or neutrality. Those trends include, among other things, the transformation of medicine into a business, the regulation of medical practice by outside interests, and the concentration of medical research on the concerns of the market. It would not be the aim of the course to counsel what to do regarding these or similar matters. Its aim, and its bias, if you will, is to make students aware that they need not be blindly submissive to so-called irresistible forces, however strong such forces may be; that they can remain their own masters, even if they decide to go along with them.

"Our time is one of rapid and incisive changes. Even the traditional goal of medicine seems to become questionable. Doctors and laymen alike used to assume that medicine should heal and keep death away as long as possible. In the days of my medical youth, complaints could be heard about how little medicine could do. I remember the cynical definition of internal medicine as *circumstare et verba facere,* 'to stand around and make words.' The situation has changed; it has become ambiguous. On the one hand, doctors and public are proud of therapeutic achievements and prolonged life expectancy, and we look forward to cures of more and more diseases. On the other hand, we reject resuscitation in cor pulmonale arrest, sign living wills, and confer medical powers of attorney. Just now I read of a poll in which nearly two-thirds of elderly people said they did not wish to become centenarians. Even in relatively good health, old age can impose burdens, financial and otherwise, on the individual, the family, and the community. Wise statesmanship, inside and outside the profession, will be required for the handling of medical matters. Wisdom cannot be taught, but history has always been considered a good school for statesmen.

"The course need not be very long, and it need not even be required. The main thing is that it be offered, so that students who do not want un-

thinkingly to follow or resist trends and demands will find encouragement in the history of their profession and uphold the proud motto of Paracelsus: He who can be his own master should not belong to somebody else (*Alterius non sit qui suus esse potest*)."

I think it advisable to end the dialogue at this point, leaving the last word to me. Even if the dean agreed that such a course should be offered, I would not be the one to give it. The course would be given by someone who would be his or her own master—and that is how it should be.

NOTES

1. See Owsei Temkin, *Hippocrates in a World of Pagans and Christians* (Baltimore: Johns Hopkins University Press, 1991), 33, 33n. 98.

2. Ibid., 61.

3. Ibid., 28–33.

4. Ibid., 254.

5. See chap. 3, this volume; in Owsei Temkin, William K. Frankena, and Sanford H. Kadish, *Respect for Life in Medicine, Philosophy, and the Law* (Baltimore: Johns Hopkins University Press, 1976), 1–23. See also Soranus, *Soranus' Gynecology*, trans. Owsei Temkin, with the assistance of Nicholson J. Eastman, Ludwig Edelstein, and Alan F. Guttmacher (Baltimore: Johns Hopkins Press, 1956; Softshell Books ed. 1991), 6.

6. Temkin, *Hippocrates*, 33.

7. Cited in Charles S. Bryan, *Osler: Inspirations from a Great Physician* (New York: Oxford University Press, 1997), 142.

8. See Bryan, ibid., 136, for examples of Osler's acting beyond duty.

9. Cited in Bryan, ibid., 137.

10. For more detail, see Owsei Temkin, *The Double Face of Janus and Other Essays in the History of Medicine* (Baltimore: Johns Hopkins University Press, 1977), 5–6.

11. See Owsei Temkin, "A Galenic model for quantitative physiological reasoning?" reprinted in *The Double Face of Janus*, 162–66.

12. From the Latin text of Celsus, *De Medicina*, proemium 23, as quoted by Edelstein, in *Ancient Medicine: Selected Papers of Ludwig Edelstein*, ed. Owsei Temkin and C. Lilian Temkin, trans. C. Lilian Temkin (Baltimore: Johns Hopkins Press, 1967), 249.

13. From the Latin text of Celsus, ibid., 26, as quoted by Edelstein, *Ancient Medicine*, 284.

14. For Edelstein's argument that vivisections did occur, see Edelstein, *Ancient Medicine*, 284 ff.

15. H. Cushing, "The binding influence of a library on a subdividing pro-

fession," *Bull Johns Hopkins Hosp* 46 (1930): 35. The passage is quoted here from Owsei Temkin, "An essay on the usefulness of medical history for medicine," reprinted in *The Double Face of Janus* (68–100), 92.

16. See chap. 3, this volume. See also Edelstein, *Ancient Medicine*, 303–16.

17. I previously mentioned this episode in *The Double Face of Janus*, 17 ff.

18. See sec. 4 of chap. 3, this volume.

19. In accepting our ability to choose and to decide as an empirical fact independent of any theories about free will, I follow a suggestion I remember from Norbert Elias in, I believe, *Die höfische Gesellschaft.*

Ethics in Medicine

CHAPTER 2

What Does the Hippocratic Oath Say?
Translation and Interpretation

The so-called Hippocratic Collection contains a large number of Greek medical writings that hail from the late fifth century B.C. to the second century A.D.[1] One of these writings carries the title, "Oath" (*horkos*). The oath does not claim an author. By the first century A.D., it was believed to be the work of Hippocrates who, by that time, had achieved legendary acclaim. There is no independent corroboration for this claim.

The oath can be divided into four main parts:

1. The performative act of swearing;
2. A covenant;
3. An ethical code, which defines:
   a. the juror's behavior to patients in general, his refusal to give poison to anybody, and his forswearing of major surgery;
   b. his behavior to patients and other inmates of a household;
4. The reward for living up to the oath and the punishment for perjury.

I now present a translation of the oath.[2] This translation and my interpretation will be broken down into the aforementioned parts. The oath is and remains a puzzling document. It is often referred to, but what the oath actually says—and what it does not say—is not always remembered. A reexamination of the content of the oath without speculations derived from other texts seemed desirable. A concluding discussion will offer a view of the document as a whole.

Translation and Interpretation of the Oath

1. I swear by Apollo Physician and by Asclepius and Hygeia and Panacea, as well as all the other gods and goddesses, making them witnesses, to fulfill this oath and this covenant, according to my power and judgment.

The juror swears by Greek healing gods as well as by other divinities. Apollo is specifically referred to in his aspect of a healer, rather than as the father of Asclepius, as the Asclepius legend has him.

The oath covers the whole document, but since the covenant, as the Greek term (*syggrapheō*) says, was in writing, the whole document was in writing.

> 2. [I swear] to hold him who has taught me this art (*technēn*) as equal to my parents, to live my life in close community with him and, if he needs money, to share with him, to deem his offspring equal to my male siblings and to teach them this art, if they wish to learn it, without a fee and without a covenant. [I swear] to pass on precepts, oral instruction, and all of the rest of learning to my own sons, to the sons of him who has taught me, and to pupils who have entered upon a covenant and bound themselves by oath to medical law (*nomōi*), and to no one else.

This is the covenant, and it will be followed by an ethical code that, I assume, stands for the "medical law" that every applicant is to swear. From this it follows that the Hippocratic Oath was a pupil's oath. This conclusion seems valid even if the Greek text were translated as "bound themselves by oath according to medical custom," instead of "medical law." This would mean, however, that the present oath was but one among possible oaths.

From the covenant we learn that the applicant was a male and that he ordinarily paid a fee to his teacher.[3] We also obtain an inkling of the substance of medical education, although we can only guess at the details. Precepts probably meant general guides for medical practice, such as the setting up of the surgery and its equipment, and perhaps also the kind of patients to be accepted and to be rejected for treatment (hopeless cases?) and arrangements regarding payment of the fee patients owed. Doctors had to make a living, and the fact that such fees are not mentioned in the oath does not allow us to assume that Hippocratic doctors treated their patients gratis. But if precepts really contained rules about the choice of patients and the fees to be charged, then such matters at the time were matters of "business," outside the ethical code. This is no more than speculation, which, however, shows how difficult it is to tell exactly what the oath did *not* take into consideration.

Oral instructions may let us think of teaching in the surgery and at the

bedside while examining and treating patients and of lectures and disputes over theoretical subjects. Even guessing is useless about "all the other learning" because we are ignorant concerning the education the pupil was expected to bring along. It may have differed from case to case. Likewise we can only guess what was meant by the "close community" of the pupil's life with his teacher. It may mean anything from the intimacy between children and parents to shared livelihood.

> 3a. I shall use dietetic regimens for the benefit of the sick in accordance with my power and judgment, as well as to protect against baneful harm and wrongdoing, nor shall I give a deadly drug to anybody if asked, nor shall I give advice leading to that end; likewise I shall not give a destructive suppository to a woman.[4] In purity and in piety,[5] I shall conduct my life and my art.
>
> I shall not cut, and certainly not sufferers from the stone, but shall yield to practitioners of this activity.

Lithotomy, cutting for stones of the bladder, was a subject of surgery that received much attention well into the nineteenth century, probably because of a greater incidence of the painful condition than now. The prohibition of lithotomy indicates that by "cutting" was meant major surgery, in which the juror was not trained and which he had to leave to practiced surgeons. Minor surgery, such as the dressing of wounds, bleeding, or the opening of abscesses, is not necessarily included in this prohibition, and the large field of dislocations and fractures is not mentioned at all. The latter is also true of pharmacology, once poisons and abortifacient suppositories are excluded. Obstetrics was in the hands of midwives, a different profession. All this suggests that, in speaking of "dietetic regimens," the oath meant not only diet in the narrow sense of food, drink, and rest but also the whole treatment and supervision of the sick person's life. The oath does not exclude teaching and practicing all aspects of the medical art (except major surgery); they pose no moral problem.

The juror swears that he will not accede to a request to give poison to anybody and will not even be a party to it. This would include murder, legal execution (preparing the hemlock?), as well as euthanasia. Likewise, he will not kill by giving a vaginal suppository deadly to the fetus to a woman who may then apply it to abort the fetus. This passage occurs in

evident parallel with the passage on poisoning. The author of the oath was surely aware that poisoning is not the only way of killing people; yet, as poisoning would be a method especially available to doctors—and doctors were the author's concern—only poisoning is mentioned. Similarly, perhaps only the "destructive pessaries" are mentioned in regard to abortion because they lay in the doctor's particular province. Whether or not the author was cognizant of other means of providing abortion and believed in their efficacy we do not know, so it is impossible to tell whether the oath prohibited abortion in general or only this particular method because it did to the fetus what a deadly poison would do to an adult.[6] In other words, the juror will not taint himself by blood guilt. He will live and practice in purity; more than that, he will conduct his life and art in a god-fearing way. A very strong promise, however interpreted.

> 3b. Into as many houses as I enter, I shall come for the benefit of the sick, without intent of any wrongdoing or baneful harm or especially of sexual acts on female or male persons, free or slave.
>
> As to what I may see or hear regarding people's life during the treatment or independently of it, regarding things that should not be divulged outside, I shall keep silence, considering such matters as shameful to be talked about.

What was said in part 3a regarding patients in general is now repeated and enlarged when the juror has entered a household with several residents. The enlargement is in two directions: sexual acts and silence regarding things that are not intended for public knowledge. Whereas modern medical ethics forbids a sexual relationship between doctor and patient only and imposes confidentiality between those two, the oath forbids a sexual relationship with any member of the household and imposes confidentiality regarding the whole household.

> 4. Now if I fulfill this oath and do not break it, may it be given to me to enjoy my life and my art, being praised among all people for all time. If, however, I transgress [against this oath] and perjure myself, may the opposite of these things take place.

The oath ends by naming the reward and punishment for living up to it and for breaking it. Both reward and punishment, even for perjury, are of this world—immortality is found in eternal praise. On the other hand,

having said repeatedly that he will do his best as far as it is in his power and judgment, the juror has protected himself against undeserved blame.

The mundane character of these expectations does not detract from the oath as a solemn undertaking. Nothing would justify the assumption that the oath is merely to present a favorable image of the healer or that the juror does not act in good faith and does not deem it right and proper to submit himself to the dictates of this oath.

Discussion

It seems strange that an oath sworn by a pupil should treat him as engaged in medical practice and teaching pupils. For a possible explanation we turn to the oath as a whole (covenant and ethical code), raising the question: why had such an oath to be sworn?

The reward and punishment incorporated in the final section will act as powerful reasons for keeping the oath, once it is sworn. But they are not a motive for swearing it in the first place. An oath will strengthen the obligations of the covenant, but the ethical code, seen as a mere resolution, could be kept without an oath. New Year's resolutions are usually not made by swearing on the Bible.

The immediate motive for swearing the oath is the teacher's insistence that the prospective pupil do so. Seen from this point of view, it also becomes understandable why covenant and ethical code are parts of the same document. The present and future obligations to his teacher and to his teacher's sons, as spelled out in the covenant, are clearly to the teacher's advantage. Moreover, stating what he will teach and the strictures upon whom he will teach establish teacher and pupil as being men who bind physicians to a high ethical code with religious overtones, such as are sworn to in the rest of the document and carry reward and punishment, like a law not to be broken.

Wherever it may be and as far as he is capable of it, the juror will treat patients exclusively for their good. He will not yield to any request to give a poison to a person, he will not even be a party to it, and he will not give a vaginal suppository killing the fetus. In short, he will live and practice his art in purity and in a god-fearing way. Also, he will leave major surgery, in which he has not been trained, to experienced surgeons. When he visits a household, the treatment of the patient will be his sole aim. He

will abstain not only from any nefarious activity but also from sexual relations with any member of the household. Neither will he repeat anything he might have learned about which silence is indicated.

The pupil might be not only the teacher's disciple but also his assistant, who will help in the patient's treatment and may be sent to the patient's house to supervise his care. It is altogether in the teacher's interest to have a pupil on whom he can rely, on whom the patient can rely, and who will not cause any trouble for the household to which he is admitted. This is the kind of man the oath represents. For the pupil, having sworn this oath may have similar advantages later in his practice. Speculating, we may even imagine that it could serve as a testimonial before a group of laymen who are examining candidates for the position of town physician.

Today, we distinguish sharply between the medical student and the doctor. Graduation marks the transformation from one stage to the other, and the Hippocratic Oath, where sworn, is usually a graduation oath.[7] However, the Hippocratic Oath says nothing about a changing status from pupil to independent practitioner. When the pupil has been accepted, he has become a healer and is treated as such.

Any interpretation has to avoid the Scylla and Charybdis of two fallacies. It must not read into the text what is not there, and it must not assume that what is not in the text does not exist. I hope that to the best of my "power and judgment" I have avoided the first fallacy; now I do not wish to be guilty of the second by assuming that what the text of the oath does not say did not exist. The text speaks of teacher and pupil, in the singular. Was that a formula for many teachers, who were all meant to use this oath? Did the oath represent a class of physicians,[8] perhaps in contrast to surgeons? Since we do not know whether this oath originally was a mere proposal or was in actual use, answers are elusive, and this is true of other questions as well. The covenant speaks of the close community between teacher and pupil. Only when we know more about the relationship between physicians and their apprentices in antiquity may we be able to obtain a clear picture of that community. The ethical code is emphatic and detailed on matters the healer must not do. It is peculiarly uncommunicative about any positive attitude toward the sick, apart from the promise of a dedicated and safe treatment. There is no word about compassion for the sick or consideration of their economic and social status.

Did such consideration not exist? Or was it left to the healer's individual initiative because his society had not yet developed a social conscience?

The foregoing interpretation has presented the oath as a sincere though pragmatic document, and this may indeed be all there is to it. But there is reason to suspect that there may be more to it. The text makes the juror's resolve to live and practice in purity and in piety appear as a climax to the preceding commitments of the ethical code. Was it a mere climax, or was it a springboard from which the preceding declarations arose? The oath does not justify itself. It does not say why it is right and proper for a healer to swear it. Was there a religious creed, a philosophical belief, or a social understanding that demanded of its medical followers this kind of oath? The text is not explicit. Nevertheless, it hints at the possibility of a religious or philosophical creed that is presupposed by the whole oath, without revealing itself in the text. And thus the Hippocratic Oath is bequeathed to us as a puzzling document.

NOTES

1. Age and its concomitants have made it impossible for me to consult libraries, including my own. I am, therefore, unable to give proper references, except to the few items listed in this chapter's bibliography. My apologies go to all authors, unknown to me or not remembered by me, who ought to have been mentioned.

2. My translation and interpretation rest on the edition of the Greek text as reprinted by Edelstein (1967), 5. For a discussion of textual traditions, see Rütten (1996, 1997).

3. Plato (see Temkin 1991, 39) speaks of "the Asclepiad Hippocrates," who, for a fee, would educate a physician.

4. Although the purpose of these vaginal suppositories (*pesson*) obviously was to effect abortion, the emphasis was on their destructiveness (i.e., directly killing the fetus). Soranus (fl. about 100 A.D.; see pp. 62–63) noted that the oath did not cover all methods of abortion mentioned by Hippocrates.

5. For a detailed analysis of this complex passage, see von Staden (1996).

6. For the debate on the oath's treatment of abortion, see Rütten (1996).

7. See the informative article by Dale C. Smith (1996).

8. If I remember rightly, Charles Lichtenthaeler suggested that the oath was sworn by the Asclepiades (see n. 3). But we are not certain whether this name denoted a definite group that considered themselves descendants of the god or whether it was a mere synonym for physicians.

REFERENCES

Edelstein, Ludwig. *Ancient Medicine: Selected Papers of Ludwig Edelstein,* ed. Owsei Temkin and C. Lilian Temkin, trans. C. Lilian Temkin. Baltimore: Johns Hopkins Press, 1967. (That volume contains, among other reprinted papers, a reprint of "The Hippocratic Oath: text, translation and interpretation").

Galvão-Sobrinho, Carlos R. "Hippocratic ideals, medical ethics, and the practice of medicine in the early Middle Ages: the legacy of the Hippocratic Oath," *J Hist Med Allied Sci* 51, no. 4 (1996): 438–55.

Lichtenthaeler, Charles. *Der Eid des Hippokrates: Ursprung und Bedeutung.* Cologne: Deutscher Ärzte-Verlag, 1984.

Rütten, Thomas. "Receptions of the Hippocratic Oath in the Renaissance: the prohibition of abortion as a case study in reception," *J Hist Med Allied Sci* 51, no. 4 (1996): 456–83.

——. "Medizinethische Themen in den deontologischen Schriften des *Corpus Hippocraticum,*" in *Médecine et Morale dans l'Antiquité: Entretiens sur l'Antiquité Classique.* Vandoeuvres-Génève: Fondation Hardt, 1997, 43:65–120.

Smith, Dale C. "The Hippocratic Oath and modern medicine," *J Hist Med Allied Sci* 51, no. 4 (1996): 484–500.

Soranus. *Soranus' Gynecology,* trans. Owsei Temkin, with the assistance of Nicholson J. Eastman, Ludwig Edelstein, and Alan F. Guttmacher. Baltimore: Johns Hopkins Press, 1956; Softshell Books ed., 1991.

Temkin, Owsei. *Hippocrates in a World of Pagans and Christians.* Baltimore: Johns Hopkins University Press, 1991.

von Staden, Heinrich. "'In a pure and holy way': personal and professional conduct in the Hippocratic Oath," *J Hist Med Allied Sci* 51, no. 4 (1996): 404–37.

The Idea of Respect for Life
in the History of Medicine

In 1948, when the horrors of the Nazi concentration camps were fresh in everybody's mind, the World Medical Association adopted what is known as the Declaration of Geneva. On admission to his profession, the young doctor was to pledge himself "to consecrate [his] life to the service of humanity." Among other pledges, he promised: "I will maintain the utmost respect for human life, from the time of conception; even under threat, I will not use my medical knowledge contrary to the laws of humanity."[1] The inhumanity of the preceding years shed light on the meaning of the laws of humanity. Now there would not be much doubt how the doctor's utmost respect for human life would direct his behavior; he would do absolutely nothing that could harm life or extinguish it. In particular, it is to be assumed that he would not practice abortion, he would not provide euthanasia, and he would not perform experiments detrimental to human beings.

The social and moral turmoil of the years since World War II has raised new debates, in the heat of which "respect for life" has tended to become a slogan. As is the way with slogans, it hides the complexity of underlying thought in order to concentrate moral force on action. Our discourse, however, will devote itself first of all to this complexity. It will not ask whether, when, and how life *should* be respected in medicine. In an analytical mood, we shall try to illuminate the content of the idea, and we shall do so by looking at some of its manifestations in the history of Western medicine, as a profession dedicated to the practice and discipline of healing.

Originally published in *Respect for Life in Medicine, Philosophy, and the Law* by Owsei Temkin, William K. Frankena, and Sanford H. Kadish (Baltimore: Johns Hopkins University Press, 1976).

Those, then, who expect a message will be disappointed, and so will historians who expect a narrative of the development of the idea through the ages. Moral ideas do not necessarily unfold with the flow of time. They have a tendency to cling to what is old, just because it is old and thereby hallowed, to what presents itself as a tradition to be invoked or refuted. We shall, therefore, start with the formation of what we shall call the *medical tradition*. Since this took place in Greece and Rome between 400 B.C. and A.D. 200, we shall, like the academic speaker of old, have to begin with "The ancient Greeks already . . ." The audience used to respond by resigning itself to forbearance, and everybody tried to make himself as comfortable as the hard chairs would allow.

We shall indeed have to come back to the ancients again and again. But considering that, fundamentally, the Geneva Declaration is an adaptation of the Hippocratic Oath and that the oath is repeated by the graduating students of many medical schools, often in the translation of the Greek text ascribed to "the father of medicine"—considering this, we shall see that in medical ethics the ancients not only preceded us but are very much with us.

In taking the Hippocratic Oath, the young physician not only promises to act for the good of his patients and to keep them from harm and injustice but he also states: "I will neither give a deadly drug to anybody if asked for it, nor will I make a suggestion to this effect. Similarly, I will not give to a woman an abortive remedy. In purity and holiness I will guard my life and my art."[2] No mention is made of the economic, social, or national status of the patients, and the physician, who is an expert in the power of drugs, forswears participation in murder, suicide, and abortion. It is generally accepted that the oath was not written by Hippocrates, who lived around 400 B.C. It did not express the common medical ethics of that time. According to Ludwig Edelstein, it reflects the thoughts of the Pythagoreans of the fourth century B.C. By the first century after Christ, it had become a well-known document, referred to by the Roman Scribonius Largus and by Soranus, a Greek practicing in Rome at the end of the century.

Soranus relates a medical controversy over the use of contraceptives and the practice of abortion: "One party banishes abortives, citing the testi-

mony of Hippocrates, who says: 'I will give to no one an abortive,' . . . because it is the specific task of medicine to guard and preserve what has been engendered by nature."[3] In referring to medicine as the guardian of life, the statement defines respect for life as a general medical responsibility. "The other party," Soranus continues, "prescribes abortives, but with discrimination, that is, they do not prescribe them when a person wishes to destroy the embryo because of adultery or out of consideration for youthful beauty; but only to prevent subsequent danger in parturition." With this latter party, which held similar views about the use of contraceptives, Soranus agreed. In modern terms, one party rejected all abortion, whereas the other party allowed therapeutic abortion. Economic reasons were not mentioned, probably because they did not exist for the rich clientele of the highly accomplished physicians whose writings have come down to us.

Scribonius Largus took the Hippocratic prohibition of abortion to be an education toward humanity. He who deemed it wrong to injure a fetus, which, after all, held only the promise of future human life, would certainly deem it even more criminal to harm a completed human being.[4]

The fame of the oath was rivaled only by that of the *Aphorisms*, another work ascribed to Hippocrates. This work begins with the well-known words: "Life is short, the art is long [*ars longa, vita brevis* in the Latin translation], opportunity fleeting, experience perilous, and the crisis difficult."[5] Here was rich food for the commentators! What did Hippocrates mean by calling experience perilous? As Galen, the great physician-philosopher of the second century, explained, the danger lay in dealing with the human body. "For in the medical art, unlike the other arts, the material is not brick, clay, wood, stone, earthenware, or hides." Those dealing with this kind of material could work without causing harm, even if they handled it badly. "But in the human body, to try out what has not been tested is not without peril, in case a bad experiment lead to the destruction of the whole organism."[6]

Let us now add an example referring to euthanasia, which is all the more instructive because it is taken from a novel, *The Golden Ass*, by Apuleius, a contemporary of Galen's.[7] It brings into relief the physician's respect for life as being above and beyond the demands of the law. A physician, we are told, was asked for a fast-acting poison, allegedly "for a sick man in the throes of an inveterate, intractable disease who longed to escape the

torture of his life,"[8] in reality for the purpose of murdering him. The physician sold a potion, but when later an innocent man was accused of murder, the physician revealed that the potion had only been a sleeping draught and not a deadly poison, "because he did not believe it proper for his calling to be instrumental in bringing death to anybody, and because he had been taught that medicine had been invented not for the destruction of man but for his welfare."[9]

If poison had really been demanded for suicide, which the Roman law did not forbid,[10] the physician would not have committed a punishable act by providing it. But apart from distrusting the buyer, the physician did not believe that his profession allowed poisoning, even for the sake of suicidal euthanasia. And, indeed, the seemingly dead man was awakened, the criminals suitably punished, and the conscientious physician rewarded.

Sufficient material has now been gathered to prove the existence of a tradition that, in its uncompromising form, did not sanction any limit to the respect for life, not even therapeutic abortion, an exception allowed by the more liberal wing. This tradition is very much with us; even the arguments used are still alive. Many physicians feel the tradition to be binding upon themselves, and many laymen expect it to be binding for the physician. Perhaps the impression has even been conveyed that we have said all there is to be said about respect for life, at least in principle, so that only details are to be added. This impression would be misleading. The tradition is ambiguous. It has had to confront other demands that influenced it, and it has not been free from contradictions within itself.

However dedicated to his profession a physician may be, he has to live in a world that makes multiple claims on him. Besides the private obligations to his family, the state is likely to make demands, if not of actual service, at least of loyalty. For this, both the forementioned Scribonius Largus and Hippocrates—not the historical Hippocrates, who lived at the time of sovereign Greek city-states in which the peripatetic physicians were strangers, but the Hippocrates of the legend that began to grow around him in Hellenistic times of strengthened common Greek patriotism— give classical examples. The legend has the king of Persia, the traditional enemy of the Greeks, look for a physician who will stop the plague decimating the Persian army. The king's agent approaches Hippocrates with

the promise of great wealth and high rank. Whereupon Hippocrates replies, "We enjoy food, garments, housing, and everything essential to life. It is not right for me to share the Persians' wealth, or to liberate from disease barbarians who are the enemies of the Greeks. Fare thee well."[11]

By the early first century, Scribonius Largus has elaborated the relationship between medicine and patriotic duty. Their profession demands that physicians have compassion and humanity, lest they be detested by gods and men. Because medicine does not measure men by their wealth or rank but promises that it will give ready help to all who implore it and will never harm anybody, a physician properly bound to medicine by an oath will not give bad medicaments, even to enemies. "But," our author says, "when circumstances [the state?] so demand, the physician as a soldier and a good citizen will pursue them [i.e., the enemies] by any means."[12]

In World War II, medical research into diseases that were of great military importance was kept secret, lest the enemy profit from it. Few researchers probably realized that they were following in the footsteps of the legendary Hippocrates, who counted his refusal to help the Persian army the equivalent of a battle won at sea.[13] Scribonius Largus apparently experienced no difficulty in reconciling his twofold obligations as physician and citizen; in his capacity of physician, he would not harm the enemy (prisoners of war?), but in his capacity of good citizen and soldier, he would fight him.[14] The potential difficulty of reconciling two different claims on the same person has been illustrated by those physicians who, during the Vietnam War, gave their Hippocratic Oath as a reason against serving in the army. Presumably they felt that, being physicians, they should be nothing but physicians and that nothing should be allowed to put limits on the respect they owed to life.

Different claims need not always be in opposition, and good citizenship could, and did, broaden the scope of respect for life. Thus the medieval doctor advised his fellow citizens—not just his patients—how to protect themselves from the plague.[15] In the early twentieth century, participation in public health, traditionally in the hands of lay administrators, could still be thought of as an obligation of the physician as a good citizen rather than as his duty as a medical man.[16]

The claims of religion, at least, we would expect always to support the medical respect for life. But even here there is no preestablished harmony. Physicians knew that a despondent patient was not in a good condition

to win the battle for life and that the physician should lift up his sagging spirits. In the words of an early medieval author, "Even as light illuminates a home and makes men see in dark shadows, so a cheerful physician turns sorrow and sadness into joy, and comforts all of the members of his patient and restores his spirits."[17] But the church, which valued the welfare of the soul more highly than the life of the body, demanded that the physician tell the patient the truth and see to it that he put his house in order and confess to the priest.[18] In obeying this demand, the physician endangered his patient's life; in disobeying, he endangered the patient's soul, as well as his own. The medieval doctor found a way out of the dilemma. He would arrange for confession before entering the sickroom. "For if this is discussed after the medical examination, the patient will begin to despair of his health, because he will believe that you also despair."[19] The French physician of the Enlightenment was advised to give an unsuspecting, mortally ill patient an inkling of his true condition, just enough "to make him aware of his duties and make him fulfill them." What those duties were was left open.[20]

The Protestant physician did not need to consider any priest, and so Thomas Percival, in his famous Code of Medical Ethics, which appeared in Manchester in 1803, declared: "The physician should be the messenger of hope and comfort to the sick; that by such cordials to the drooping spirit, he may smooth the bed of death, revive expiring life, and counteract the depressing influence of those maladies which rob the physician of fortitude, and the Christian of consolation."[21] Percival's English Code formed the basis of the code of ethics of 1847 of the American Medical Association. The whole passage was reprinted except for the end, which now read "those maladies which often disturb the tranquillity of the most resigned in their last moments."[22] This could be acceptable to Catholics, Protestants, Jews, and freethinkers alike and was proper for a society in which state and religion were separated.

Through all these examples the medical tradition takes death to be the ultimate enemy. Dying is opposed to the very last. The doctor encourages the patient's will to live, for to prepare him for death is not his medical duty. He must, of course, be tactful. The ancient physician who had cited to his despairing patient the Homeric line (*Iliad* 21.107): "Patroclus also died, and he was much better than you," was held up as a horrid example of how not to act.[23] There remained the vexing problem of whether the

patient ought to be told the truth, or whether this should be left to his relatives. But apart from tact and the silent respect paid to the dying, the *ars moriendi,* the art of dying, was in the hands of ministers and philosophers rather than physicians.

Terminal illness has become a special medical problem for us today. We are more aware of dying as a process inherent in the process of life; we speak of a death instinct in man, and the beginning and ending of life have become medical and legal issues.

➤➤ • ◄◄

The state and the church, or social and religious life, have appeared before us as historical forces that molded the contours of respect for life in medicine. To speak of contours is to use a spatial metaphor—the medical tradition is inside and the other forces impinge on it from outside—that is somewhat simplistic, but it is analytically useful and allows us to turn now to the contradictions *within* the medical tradition.

As an example we cite the case of the infant that cannot pass through the birth canal, be it that the mother's pelvis is too narrow, be it that hydrocephalus has pathologically enlarged the infant's head, or be it for any other cause. Up to about a hundred years ago, the physician was faced with a cruel decision because the life of the mother competed with that of the infant. Cesarean section, which might save the infant, spelled almost certain death for the mother or, from about the eighteenth century to the later nineteenth, involved at least a very high risk. Should he sacrifice the mother for the slim chance of saving the child? Should he save the mother at the expense of the infant, which, if not already dead, would be killed in the process of diminishing its size? Which life demanded more respect? Or was there a higher law that forbade the sacrifice of either and thereby all but ensured the death of both?

It might be argued that the decision was already made when the physician decided for or against abortion and that Soranus, who admitted performing therapeutic abortion, was consistent in teaching the destruction of the infant, with the explanation that, "even if one loses the infant, it is still necessary to take care of the mother."[24] But some early Christian authors saw the matter in a different light. Tertullian, the great patristic writer of the third century, decried abortion as murder, yet he condoned killing the infant, lest it become "a murderer of the mother."[25] I am un-

35

able to say how far the voice of Tertullian and the similar voices of Augustine and Eusebius, countenancing destruction of the unborn child as a last resort, were heeded or whether they were lost in the general theological condemnation of abortion and surgical destruction (craniotomy, insertion of hooks, embryotomy) said to have prevailed during the Middle Ages.[26]

The anguish of the physician is well described by an author of the late sixteenth century. In a case of difficult labor and in doubt whether the child was still alive or dead, "What should Christian physicians do and what plan should they follow? For if they abandon the patient without help, they will be accused of being inhuman; if, however, they rashly have recourse to manual operation and to pulling with instruments and, to save the woman, kill the fetus that, perchance still alive, has not yet been baptized, they will burden their own conscience, since evil things should never be done that good may result. They will have to account for their deed to the Highest." Our author then describes what he does when, "upon the urging of Christian piety," though "somewhat reluctantly," he has to deal with such a case. If everything seems all right, "having implored the help of almighty God, I begin the work with alacrity and usually bring it to a successful end. But if the female parts . . . show considerable narrowness, I deem it better to take honest flight and to refrain from the task than to take such grave risks, all the more if the woman is of a weak constitution."[27]

In the early nineteenth century, British obstetricians charged that cesarean section was performed more frequently on the Catholic Continent than in England because of the willingness to risk the mother's life rather than leave the infant unbaptized.[28] But in England, where the sacrifice of the infant was deemed preferable, the decision pro or con was equally based on moral rather than strictly medical grounds. "Where are the circumstances," an opponent of cesarean section asked, "that can ever warrant the certain endangerment, nay, often the more probable sacrifice, of a mother's life for the chance—and be it remembered it is nothing more—of preserving that of her child? How few of the children that 'have been ripped from their mother's belly like the Thane of Cawdor' have been reared?"[29] But the alternative—namely, killing the child—was condemned with equally strong and plastic arguments. "Assuredly no man would consider himself justified, on any plea whatsoever, in perforating, and breaking down with a pointed iron instrument, the skull of a living child an

hour after birth, and subsequently scooping out its brain. But is the crime less when perpetrated an hour before birth?"[30]

Before the advent of safe cesarean section and other very modern techniques, the dilemma was not solvable from inside the medical profession. The obstetrician who was an opponent of abortion might yet feel compelled to save his patient, rather than watch her die. In its elementary form the dilemma was inherent in the Hippocratic Oath, which forbade the physician to give a remedy causing the premature death of the fetus but also charged him with the welfare of the patient—that is, the mother.

"Respect for life" has proved ambiguous in the history of medicine as practice. It proves no less ambiguous in the history of medicine as discipline and in the history of the relationship between the two. The practice of medicine rests on respect for the patient, or a particular group of patients. The physician is responsible for this particular life, or these particular lives. Since all healing is based on some kind of knowledge, the cultivation of such knowledge is part of medicine, and thus human anatomy, physiology, and pathology, as well as clinical observation, are *medical* sciences. As a scientist, the physician also deals with human life, but in a general, abstract form, not with the particular life of this or that patient. The medical scientist determines the boundaries between life and death. In our time, this task has led to important redefinitions.

Long ago, the heart was defined as the *ultimum moriens,* the organ that died last. Early in our century, experiments showed that the isolated and refrigerated hearts of mammals could be revived after nearly two days.[31] The brain showed no such endurance, though its death, too, entailed the death of the whole organism. If the brain was made the indicator of life or death, a person in whom it had ceased to function might be pronounced dead and the as-yet-living heart be transplanted to another individual. This is a well-known, dramatic example of how science redefines human life and death biologically, leaving the legal definition to the lawyer. It shows the near-miraculous achievements of medical science within the last hundred years, since the recognition of Louis Pasteur's germ theory of infectious disease and Lister's application of it in surgery. Pasteur was a chemist, not a physician, and since his time many scientists without medical degrees have worked in the field of medical research. The researcher

for a serum or a drug that will eradicate a disease does not have to think of sick individuals. The geneticist who cautions against certain techniques of constructing DNA (deoxyribonucleic acid) molecules, the carriers of inheritance, has in mind dangers that threaten mankind.[32]

All this sounds very modern, and so it is, as long as our attention rests on scientific progress and technical achievement. But, the underlying principle is very old. When medical scientists who lived in the third century B.C. were blamed for having conducted their research on criminals condemned to die, their followers defended them and maintained that it was not cruel "that in the execution of criminals, and but a few of them, we should seek remedies for innocent people of all future ages."[33] In other words, the physician's respect for life and the scientist's respect for human life in general can clash.

Since the middle of the last century, medical research and science have inclined toward the exact sciences, so that heredity is seen as a chapter in the chemistry of macromolecules, the brain as the likeness of a computer, and metabolism as a series of enzyme reactions. The old simile of the body as a machine seems irresistible, and many biologists and physicians think of life as reducible to the laws of physics and chemistry. A good physician must have good scientific training, yet what kind of respect can he be expected to have for the life of his patient if he sees in him nothing but a machine? Is the respect that a physician owes his patient the same kind of respect a mechanic may have for an expensive car? The question has been denied so often, or rather, becoming a "mere" mechanic has been held up so often as an undesirable development for the physician, that we can take a negative answer for granted. The kind of respect we have depends on the subject we are dealing with, and if we deny that man and machine demand the same respect, we imply a difference between man and machine, even if we accept reductionism in biology. The physician who has to alleviate pain deals with a feeling machine, and he must take into account that, not only should this machine function in human society, but also it can express its wishes and decide how it wants to function. The auto mechanic will consult the wishes of the owner of the car. But man owns himself; the physician must be guided by what the patient can do and wants to do in a social, cultural, and intellectual milieu.

Our analysis will sound less abstract if we translate it into the language of health and disease. Disease is undesirable: it is "bad life." There are de-

grees in the severity of disease, from impairments of *joie de vivre* and social functioning, through an impairment of vegetative and animal functions, to a threat to life itself. The evaluation of a disease goes beyond biology. A stroke incapacitating a man's power of movement can have greater consequences in a society of peasants and craftsmen than in a society of white-collar workers. On the other hand, a stroke that leads to dyslexia, a difficulty with reading, is disastrous for an intellectual, whereas it may go unnoticed among illiterates. Biologically speaking, both are lesions of the brain, varying in extent and localization. Health—or "good life"—also has its gradations, from excellent to bearable, and mere absence of disease is no longer considered synonymous with health. The World Health Organization has gone so far as to identify health with "a state of complete physical, mental, and social well-being."[34] Surely, this extends beyond biological values.

It was a physician, La Mettrie, who insisted that man was nothing but a machine. But as long as the physician remembers the situation of the human being for whom he is responsible, his metabiological beliefs are of secondary importance. Responsibility in practical medicine requires both scientific detachment and a sense of duty to maintain the patient in the best possible condition.

Having allowed us to proceed so far, the devil's advocate may no longer be able to restrain himself. "You speak of respect for life in medicine, of responsibility, and of duty," he says. "The layman, the patient, expects them, but the doctors do not live up to these expectations. You have bored us with antiquity; well, let me quote Martial, who said of an eye doctor turned gladiator: 'As a physician you did what you are now doing as a gladiator!'[35] Surely, a gladiator is not a respecter of life! You have talked of the pious medieval doctor; well, the Middle Ages saw through him and coined the adage: *'Dat Galenus opes'* [Galen gives riches]. The soaring costs of medical care speak a clear language. And as to modern times, Molière unmasked the arrogance of a profession that thinks to rule over life and death. What did Dr. Purgon tell the patient when his authority was questioned?"

> I have to tell you that I abandon you to your bad constitution, to the distemper of your bowels, to the corruption of your blood, to the acrimony

of your bile, and to the feculence of your humors. And I wish that before four days have passed you may be in an incurable state . . . that you fall sick of bradypepsia . . . go from bradypepsia to dyspepsia . . . from dyspepsia to apepsia . . . from apepsia to lientery . . . from lientery to dysentery . . . from dysentery to dropsy . . . from dropsy to the loss of life to which your folly has led you![36]

"These are the true, unconscious thoughts of the doctors! Sigmund Freud could not have expressed it better—if only he had not been a doctor himself!"

Get thee hence, Satan! Such scurrilous remarks must not be dignified by any reply! Like most critics, the devil's advocate attacks physicians, not their art and science. Medicine can as little be judged by the behavior of some of its least admirable adepts as the Christian, Jewish, and Mohammedan religions can be judged by the behavior of some of the least admirable of their faithful followers.

But the devil's advocate is not silenced quite so easily. "If medicine really had respect for life," he continues, "it would demand that all life be respected, not only that of this or that patient or of mankind in the abstract. It would demand that the death penalty be abolished, that all wars be ended, and that poverty cease to be a cause of disease and of death and hinder the patient's complete physical, mental, and social well-being. The defense that individual physicians must not be mistaken for medicine will not protect us here. On the contrary, some physicians have opposed the death penalty, have been pacifists, and have striven for social reform. But medicine as an art and science has not."

What have we to say to all this?

→> • <←

Medicine does not deal with life per se; this is the biologist's concern. To physicians, "respect for life" has meant "respect for human life." Medicine has not stood for vegetarianism, and animals have been sacrificed in experiments aimed at saving or improving human life. Respect for life in medicine has, moreover, revealed itself as a complex and paradoxical idea, not free from contradictions. Medicine is silent about the meaning of the life for which it claims respect. Strictly speaking, medicine is not a person and can neither stand for something nor claim anything. "Standing

for" and "claiming" are no more than convenient abbreviations, helpful in exploring how far respect for life is necessary in the practice and discipline of healing.

The meaning of life does not have to enter into medical thought. The separation of medicine and philosophy has been attributed to Hippocrates. Before him, our ancient source relates, "the curing of diseases and the contemplation of the nature of things" were in the same hands, because medical knowledge was needed, "especially by those whose bodily strength had become weakened by quiet thinking and watching by night."[37] Medicine is very much concerned with whether a patient can live or will have to die. But in its metaphysical sense, "to be or not to be" is not a medical question. In dealing with suicide, the medical man is likely to ask what made life seem worthless; he is not likely to ponder the value of life. The paradoxical nature of the situation has revealed itself in the problem of euthanasia. Even if the patient's life is reduced to a state of vegetation, medical tradition—and, since the Middle Ages, the law as well—binds the medical man to respect this life. But medical tradition does not tell him *why* such a life should be protected and maintained, nor does it make the life of a saint more worthy of respect than the life of a sinner.

Medical respect for life exists in a void, and human nature, as well as external nature, abhors a void. Historically speaking, the void has been filled in two different ways. The one way we already encountered in various examples, in which demands pressed on medicine from outside. What presses from outside can also be an answer requested from inside on grounds of religion, philosophy, or sociopolitical ideologies. Even the Hippocratic Oath probably had a religious and philosophical background. Then Stoic ideas of philanthropy entered medicine.[38] The monotheistic religions made charity a duty and made abortion, suicide, and euthanasia sins. God had given life, and man must not interfere with His purpose. In more modern times, work has been seen as a social and religious duty— witness Paracelsus, the contemporary of Luther, who urged physicians to search incessantly for remedies to cure diseases deemed incurable. Paracelsus's fame rested largely on his use of chemical processes to extract from drugs the potencies active against disease.[39] This chemotherapeutic notion, developed by Ehrlich into the concept of "magic bullets" killing infectious organisms, is still popular, and it fits in with our emphasis on

work as necessity and duty. Potent drugs can attack man's disease, enabling him to carry on with his work and habit of life. There are even those who let cocktails and tranquilizers take care of the stress of a life that is allowed to remain as stressful as before. Life has many headaches. Take aspirin, and life will be free of all headaches!

With the awakening of the Western social conscience toward the end of the eighteenth century, not only were ideas of social reform accepted, but they were often even advocated by physicians. The revolutionary period of 1848 was a culminating point. As Erwin H. Ackerknecht showed, medical science was believed to be the science that yielded the knowledge of man fundamental for all social sciences, and the physician was the natural advocate of the poor.[40] After the revolutionary fever subsided, these ideas lost much of their hold on physicians, but they have come back to the fore again today in discussions over health insurance and the promotion of health. Ideas of equality that spring from political and social ideologies thus urge that respect for life must be equal respect for everybody's life. If accepted or rejected with sufficient force, these ideas make the physician turn to politics.

The other way of filling the void is to make health and life the *summum bonum*, not by explicit philosophy, but by letting respect for life operate without regard for other values that make life only a means to different ends. Health became a goal in itself in ancient, so-called dietetic medicine, which regulated man's entire life.[41] Health was thought of as a precarious balance of the body constituents. The slightest mistake in food and drink, exercise and rest, even in the sphere of emotions, could lead to illness. A man rich enough and with sufficient leisure could protect himself by perpetual medical surveillance. Of course, he had to lead the life of a hypochondriac, hypochondria being a disease to which those who live for the sake of their health are prone.

In its pure form, dietetic medicine flourished in Hellenistic Greece. The Romans had other things to do than to give themselves up to a life dedicated to their health; they were reminded by Vergil that through their empire they were supposed to rule the nations. But dietetic medicine is more than a historical episode. It represents an idea ever-inherent in medicine; the temptation is great to divide mankind into two categories of people: physicians and patients. And physicians are not the only ones to be tempted to do this. In reality, most of us live in a state of compromise.

We believe in values beyond health, yet we also sometimes forget them and respect our lives more than anything else. Physicians and patients know how difficult it is to find the proper compromise.

"To be or not to be" is not a medical question, because once a person has entered medicine, he or she has already voted for being and has assumed respect for life as a responsibility and a duty. But different people have chosen the profession of medicine for different reasons, some lofty and some not so lofty. Respect for life is a necessary condition for being a physician; it is not a necessary motive for entering medicine. The question of how deeply respect for life must be felt, as long as responsibility and duty are not shirked, must be left open. Isn't this an austere, formalistic concept of respect for life in medicine, closer to Kant's categorical imperative than to the pulsating life and warm devotion of many physicians? Albert Schweitzer taught and showed what reverence for life could and should be. Indeed, *Ehrfurcht vor dem Leben* became a focal point of Schweitzer's philosophy. "The ethic of reverence for life," he wrote, "thus includes all that can be designated as love, devotion, compassion, sharing of joy and of aspirations."[42] Reverence (*Ehrfurcht*) is more than respect (*Achtung*). Whatever the semantic subtleties, we must not forget that Schweitzer was a theologian and a philosopher before he added medicine to his qualifications. Those ideas came to him in his striving for a philosophy of civilization, though they may, of course, have been reinforced by his medical activities in Africa. The philosopher will ponder the ethics of respect for life and its manifold meanings, and his speculations will go beyond its role in medicine.

Lest we be disappointed with the merely regulative nature of the idea in medicine, let us remember that respect for life does not exhaust the ethical foundation of medicine. Healing in the broadest sense goes beyond respect for life. Scribonius Largus said that physicians lacking in humanity and compassion were detested by men and gods. Paracelsus warned doctors that they would fail if they lacked compassion and love.

If I may briefly step out of my role as historian, I would like to hope that a good physician will not only show respect for life but also feel it. Otherwise, he may not only easily become entangled in such snares as any inhuman regime may have ready for him but may fail, under any regime, to satisfy the human needs of the sick.

NOTES

1. *World Medical Journal* 11 (1964): 356G. The declaration can also be found in many earlier volumes of that journal.

2. Quoted from the English translation by Ludwig Edelstein, *The Hippocratic Oath*, reprinted in *Ancient Medicine: Selected Papers of Ludwig Edelstein*, ed. Owsei Temkin and C. Lilian Temkin (Baltimore: Johns Hopkins Press, 1967), 6.

3. *Soranus' Gynecology*, trans. Owsei Temkin with the assistance of Nicholson J. Eastman, Ludwig Edelstein, and Alan F. Guttmacher (Baltimore: Johns Hopkins Press, 1956), 63 (bk. 1, chap. 19).

4. Scribonius Largus, *Conpositiones*, ed. Georg Helmreich (Leipzig: Teubner, 1887), preface (2,27–3,2). In Karl Deichgräber's edition of the preface, *Professio medici: Zum Vorwort des Scribonius Largus* (Wiesbaden: Franz Steiner, 1950, for Akademie der Wissenschaften und der Literatur, Mainz [Abhandlungen der geistes- und sozialwissenschaftlichen Klasse, 1950, no. 9], 875–79), the text of this passage is identical.

5. For the Greek text, see W. H. S. Jones's edition of *Hippocrates*, 4:98, in the Loeb Classical Library. My translation differs somewhat from that of Jones.

6. Galen, *In Hippocratis Aphorismos commentarii* 1 in *Opera omnia*, ed. Carl Gottlob Kühn (Leipzig: Cnobloch, 1821–33), 17B:353, 12–354,4.

7. Apuleius, *Metamorphoses* 10.9–12. I have used *Apuleius, Metamorphosen oder Der goldene Esel*, Latin and German by Rudolf Helm, 6th ed., by Werner Krenkel (Berlin: Akademie-Verlag, 1970), 330–33.

8. *Metamorphoses* 10.9, p. 330,4–6. Helm, 331, translates the difficult *veterno* by *langwierig*.

9. *Metamorphoses* 10.11, 330,34–332,1.

10. Darrel W. Amundsen, "Romanticizing the ancient medical profession: the characterization of the physician in the Graeco-Roman novel," *Bull Hist Med* 48 (1974): 320–37 (see 323, 325).

11. *Epistulae* 5, ed. E. Littré, *Oeuvres complètes d'Hippocrate* 9 (Paris: Baillière et Fils, 1861), 316–18. In spite of the word *themis*, I cannot follow Deichgräber, *Professio medici*, 857, who sees a religious element in Hippocrates's refusal. The letters indicate that, apart from his patriotic reasons, Hippocrates objects to the imputation of being mercenary. For Hippocrates's patriotism, see Edelstein, "Hippokrates," in Pauly-Wissowa, *Real-Encyclopädie*, Suppl 6 (1935): cols. 1300–1301.

12. Even where not put in quotation marks, the paraphrase follows the wording of the text rather closely. Deichgräber's edition of the passage, p. 876,29–37, which I have accepted, differs slightly from that of Helmreich (see above, n. 4), 2,16–26, but it seems to correspond more closely to my interpretation, which I

have tried to bring out by rearranging the sequence of the arguments. See also Deichgräber, *Professio medici*, 867. Edelstein, "The professional ethics of the Greek physician," reprinted in *Ancient Medicine*, 319–48, on p. 339 translates the passage *qui sacramento medicinae legitime est obligatus* (ed. Helmreich, 2,20–21; Deichgräber, 876,32–33) "bound in lawful obedience to medicine by his military oath." The assumption of a special oath for army surgeons seems unwarranted to me.

13. *Epistulae* 11, ed. Littré, 9:328,14.

14. Cf. Edelstein, "The professional ethics," 340, 342 and Deichgräber, *Professio medici*, 867. Scribonius, in speaking of compassion (*misericordia*) and humanity (*humanitas*), states the motives that lead him to respect the life of friend and foe alike. The Hippocratic Oath makes no reference to wealth and rank of patients, and the Hippocratic writings admonish the physician "to help, or at least to do no harm" (*Epidemics* 1.11; Jones's translation, 1:165, of the Loeb edition), which corresponds to Scribonius Largus, ed. Helmreich, 3,5 (ed. Deichgräber, 876,44 f.): "for medicine is the science of healing, not of harming." The whole complex of questions here discussed has been commented on by Fridolf Kudlien, "Medical ethics and popular ethics in Greece and Rome," *Clio medica* 5 (1970): 91–121 (esp. 91–96). I cannot, however, agree that the dilemma between physician and good citizen existed for the peripatetic Hippocratic physician as it existed for the Roman Scribonius Largus.

15. For instance, Jacme d'Agramont at the beginning of his *Regiment de preservacio a epidemia opestilencia e mortaldats*, trans. M. L. Duran-Reynals and C.-E. A. Winslow, *Bull Hist Med* 23 (1949): 57–89.

16. "The Principles of Medical Ethics of the American Medical Association" (adopted in 1912), chap. 3, sec. 1: "Physicians, as good citizens and because their professional training specially qualifies them to render their service, should give advice concerning the public health of the community," quoted from the reprint in Chauncey D. Leake, ed., *Percival's Medical Ethics* (Baltimore: Williams and Wilkins, 1927), 269.

17. The passage occurs in the letter of Arsenius to Nepotian and is here quoted from Loren C. MacKinney, "Medical ethics and etiquette in the early middle ages: the persistence of Hippocratic ideals," *Bull Hist Med* 26 (1952): 1–31 (see 12).

18. See Paul Diepgen, *Die Theologie und der ärztliche Stand* (Berlin-Grunewald: Walter Rothschild, 1922), 48 ff.

19. *De adventu medici ad aegrotum*, in Salvatore De Renzi, *Collectio Salernitana* 2 (Naples: Filiatre-Sebezio, 1853), 74–80 (see 74). See also Diepgen, *Die Theologie*.

20. Article "Prognostic" in *Encyclopédie ou Dictionnaire raisonné des sciences, des arts et des métiers* (new ed.), 27 (Geneva: Pellet, 1778): 524–25 (see 524).

21. Percival, *Medical Ethics*, chap. 2, art. 3; quoted from the reprint by Leake, *Percival's Medical Ethics*, 91.

22. "Code of Ethics of the American Medical Association" (as adapted 1847 and published 1848), chap. 1, art. 4; quoted from the reprint by Leake, *Percival's Medical Ethics*, 220 ff.

23. Galen, *In Hippocratis Epidemiarum Commentaria* 4.10; ed. Ernst Wenkebach and Franz Pfaff, *Corpus Medicorum Graecorum* 5.10,2,2; reprinted Berlin: Academia Litterarum, 1956, 203. For commentary see Karl Deichgräber, *Medicus gratiosus: Untersuchung zu einem griechischen Arztbild* (Wiesbaden: Franz Steiner, 1970, for Akademie der Wissenschaften und der Literatur, Mainz [Abhandlungen der geistes- und sozialwissenschaftlichen Klasse, 1970, no. 3], 33 (223).

24. *Soranus' Gynecology*, 189 (bk. 4, chap. 3, art. 9).

25. Tertullian, *De anima* 25.4–6; ed. J. H. Waszink (Amsterdam: Meulenhoff, 1947), 36, 326. Cf. Franz Joseph Dölger, "Das Lebensrecht des ungeborenen Kindes und die Fruchtabtreibung in der Bewertung der heidnischen und christlichen Antike," *Antike und Christentum* 4 (1934): 1–61 (see 42, 49); also Ruth Hähnel, "Der künstliche Abortus im Altertum," *Sudhoffs Arch* 29 (1937): 224–55.

26. For Augustine and Eusebius see Dölger, "Das Lebensrecht," 44–49, 280–81. Tertullian's reluctant acceptance of surgical destruction rested on the human right (here the mother's) to turn against anybody threatening one's God-given life, a principle for which later theologians (see below, n. 28) cited Saint Thomas Aquinas "and most theologians." It should, however, not be overlooked that Tertullian's excuse of the physician's practice is merely part of his argument for the fetus's possessing a soul. In the question of destruction, as well as of baptism in the womb, much casuistry seems to have been involved, for which nn. 27 and 28 cite examples and for which articles in *The Casuist* (New York: Wagner, 1906 and 1910) 1:331–39, 3:178–81 (acquaintance with which I owe to the courtesy of Stephen M. Winters) may also be consulted.

27. Julius Caesar Arantius, *Anatomicae observationes*, chap. 39 in *De humano foetu liber*, 3d ed. (Venice: Jacob Brechtanus, 1587), 106 ff. Cf. Heinrich Fasbender, *Geschichte der Geburtshilfe* (Jena: Gustav Fischer, 1906), 111 ff. François Mauriceau (*Traité des maladies des femmes grosses et de celles qui sont accouchées*, 4th ed. [Paris: Laurent d'Houry, 1694], 295 ff.) seems to have been one of the earliest modern obstetricians to admit practicing embryotomy on a living child, if it prevented the mother's certain death. (See also Fasbender, 172 n. 1.) Mauriceau's English translator, Hugh Chamberlen (*The Diseases of Women with Child and in Child-Bed* [London: Bell, 1710]), remarks in his preface (xv): "In the 17th chapter of the second Book, my Author justifies the fastning Hooks in the Head of a Child that comes right, and yet because of some Difficulty or Disproportion cannot pass; which

I confess has been and is yet the Practice of most expert Artists in Midwifery not only in England, but throughout Europe; and has much caused the Report, That where a Man [i.e., an obstetrician] comes, one or both must necessarily die; and is the reason for forbearing to send, till the Child is dead, or the Mother dying." This suggests that killing the child to save the mother was a more widespread practice than the literature indicates; see also below, n. 28.

28. J. H. Young, *Caesarean Section: The History and Development of the Operation from Earliest Times* (London: Lewis and Co., 1944), 90 ff. In this connection (cf. 39–40), an opinion by the theologians of the Sorbonne on the advisability of cesarean section is of interest. The matter is to be found in the French translation (with additions) by Jacques-Jean Bruier d'Ablaincourt of Deventer's work, in which craniotomy was taught. This French edition (*Observations importantes sur le manuel des accouchemens, traduite du Latin de M. Henry de Deventer,* première partie [Paris: Giffart, 1734]) contains a chapter on cesarean section (345–57) followed by a memoir requesting an opinion from the doctors of theology of the University of Paris (357–59) regarding the alternative of embryotomy vs. cesarean section, together with the possibility of baptizing the unborn child. The reply (359–65) focused on cesarean section, so that the alternative, embryotomy, was implied rather than treated explicitly. Cesarean section was to be performed if there was hope of saving both mother and child. It was not allowed if it would lead to the mother's certain death, for this would be homicide, and it was not permissible "to do evil in order to achieve good" (i.e., the infant's baptism). Cesarean section was allowed, even if its success was doubtful, if otherwise both mother and child faced certain death. If, however, the operation would save only mother or child without well-founded hope for both, then mere justice allowed the "use of every proper means" to save the mother's life, even "exposing the child to certain death." However, "charity demands that the mother prefer the salvation of her child to her own life, if the infant's baptism can be procured only at the expense of her death." Quite apart from the insistence that the mother choose charity, did "every proper means" imply embryotomy (including craniotomy)?

The Sorbonne was reasonably clear on cesarean section, but not equally so on embryotomy of the living child, be it that it was taken for granted, be it that a clear answer was shunned. Nevertheless, this decision of 1733 did not categorically exclude the practice of Mauriceau and other surgeons. The Sorbonne viewed the chances of successful cesarean section relatively optimistically. The theological attitude against embryotomy hardened in the nineteenth century, when the chances of a woman surviving a cesarean section greatly improved (see "abortion," in *The Catholic Encyclopedia* 1 [New York: Universal Knowledge Foundation, 1907], 47 ff., and the *Casuist* 3:178–81).

29. Anonymous reviewer of 1843, cited by Young, *Caesarean Section*, 79. The reviewer claimed that cesarean section was frequently performed on the Continent because the patients there (especially in France), "at least those amongst the poorer classes, seem to be regarded, not so much as fellow creatures that have the same hopes and desires as ourselves, but rather as objects, so to speak, of natural history, which the learned doctor has to speculate and experiment upon." C. Lilian Temkin drew my attention to a mistake in the quotation from Shakespeare's *Macbeth* (act 5, scene 7): Not the Thane of Cawdor—i.e., Macbeth—but "Macduff was from his mother's womb untimely ripp'd."

30. *The Obstetric Memoirs and Contributions of James Y. Simpson*, ed. W. O. Priestley and Horatio R. Storer, 2 vols. (Philadelphia: Lippincott, 1855–56), 1:540. The section of the article in which the quoted passage appears had previously been published in the *Edinburgh Monthly Journal of Medical Science*, 1852, and was, therefore, not intended for laymen.

31. A. Kuliabko, "Studien über die Wiederbelebung des Herzens," *Pflügers Arch* 90 (1902): 461–71.

32. See the report of Nicholas Wade in *Science* 187 (1975): 931–35.

33. Celsus, *De medicina*, prooemium 26; trans. W. G. Spencer (Loeb Classical Library), 1:15.

34. "Constitution of the World Health Organization," *Chronicle WHO* 1 (1947): 29.

35. Martial, *Epigrams* 8.74.

36. *Le malade imaginaire*, act 3, scene 5. The interjections by the other people in the scene have been omitted.

37. Celsus, *De medicina*, prooemium 6–7; Spencer's trans., 1:5.

38. See Edelstein, "The professional ethics," 329 ff.

39. See Owsei Temkin, *The Falling Sickness*, 2d ed. (Baltimore: Johns Hopkins Press, 1971), 170–72.

40. Erwin H. Ackerknecht, *Rudolf Virchow: Doctor, Statesman, Anthropologist* (Madison: Univ. Wisconsin Press, 1953), 44–46.

41. This discussion of ancient dietetics is based on Ludwig Edelstein, "The dietetics of antiquity," in *Ancient Medicine*, 303–16. This article originally appeared in German in 1931.

42. Albert Schweitzer, *Aus meinem Leben und Denken* (Hamburg: Richard Meiner, 1954), 134. Cf. the English translation by C. T. Campion, *Out of My Life and Thought* (New York: Holt, 1949), 159.

CHAPTER 4

Some Moral Implications
of the Concept of Disease

A leading American medical dictionary defines pathology, literally the study of disease, as "that branch of medicine which treats of the *essential nature* [italics added] of disease, especially of the structural and functional changes in tissues and organs of the body which cause or are caused by disease."[1] "Essential nature" means that which underlies disease, which is inseparable from it and not accidental like changing subjective symptoms, idiosyncrasies, or incidental agents as different from etiological factors, such as microorganisms. The allegedly "essential nature" of disease does not include any moral stigmata attached to it. Works on pathological anatomy and pathological physiology study the "abnormal" structure and functions of the organism, just as works on "normal" anatomy and physiology study structure and function in the state of health.

Ideally, these studies are supposed to use scientific, objective methods free from value judgments (*wertfrei*). How far this ideal is realized is another matter; all kinds of biases as to race, sex, class, and so on, may enter.[2] But watchful criticism is supposed to weed them out. This task may be far from easy, especially in the case of mental disease. For instance, it took a long time to separate alcoholism (a term introduced only in the nineteenth century) as a disease from drunkenness as a vice engendered by a corrupt act of will.[3] Benjamin Rush sketched "a moral and physical thermometer," where the highest degree of temperance, drinking only water, entailed "health and wealth" and the lowest degree of intemperance, drinking gin, brandy, and rum day and night, led to "murder," "madness and despair,"

Originally published in *Festschrift für Erna Lesky zum 70. Geburtstag*, edited by Kurt Ganzinger, Manfred Skopec, and Helmut Wyklicky (Vienna: Brüder Hollinek, 1981).

and the "gallows."[4] Hysteria was treated as a kind of depravity even much later and, in our own century, the shellshocked soldier was often viewed as a coward and shirker.

Once vice has been transformed into disease, it seems but logical to let crime and sin follow suit on the assumption that free choice does not exist, that they also need treatment rather than punishment. For it is taken for granted that, where disease begins, free will ends, and that the "essential nature" of disease rests in processes that are determined physically or psychologically.

→→ • ←←

The history of syphilis illustrates how the "essential nature" of a disease can be seen as independent from the moral attributes attached to it.[5] The name *syphilis* denotes the disease that we attribute to an infection with *Treponema pallidum* and that has been recognized as a clinical entity since 1494, perhaps even 1493. We can describe the history of "the disease itself," that is, its varying anatomical and clinical manifestations over the centuries, and we can describe the development of our knowledge of the disease and its cause. Third, we can review the changes in the moral assessment of the disease under varying social conditions. Until its venereal character was generally recognized (about 1520), it was taken as a divine punishment for the sinful life of mankind. Then it became bound to the sexual morality of the times. To the common folk it was a punishment for sin—but now of personal sin. For the nobility, setting the tone at court and in the army, it was a regrettable incident in the life of a man of the world, it was a cavalier's disease; there just were no roses without thorns. Even the treatment tended to differ: the higher strata of society in the sixteenth century preferred a decoction of guaiacum, prescribed by many learned physicians, over the inunction of mercury with which barber surgeons treated the disease to the point of severe mercurial poisoning. With the rise of an "enlightened" middle class in the eighteenth century, for whom family life was sacred, syphilis became a shameful disease. Sexual purity was fostered among the young by showing them in hospital wards the terrible consequences of vice. The growing social consciousness of the later nineteenth century stressed the danger of syphilis, not only for the family but for the health of the nation and of society at large. Prophylactic measures were promoted, and a German law of the 1920s enforced the treat-

ment of syphilis and made sexual intercourse during the infectious period a criminal offense.

The tripartite distinction of "the disease itself," the knowledge of it, and its evaluation in human society is useful and necessary. In reality, however, such a neat division does not exist. A "natural history of disease," in the strict sense of complete independence from culture, is a fiction.[6]

As long as disease is an object of scientific study, the inability to find a physical cause for it, or the unwillingness to look beyond physical causes, makes its occurrence something of an accident (i.e., outside its "essential nature"). Disease as accident was a liberating concept for rising scientific medicine. It permitted the pathologist to concentrate on "the structural and functional changes in tissues and organs." Accepting Henle's definition of disease as life under changed circumstances and intent on investigating these circumstances and their consequences in the body, the young Virchow in 1847 wrote: "Es ist nun einmal kein Zweck darin zu entdecken, wenn einer eine Geschwulst bekommt: es ist, wie wir zu sagen pflegen, ein Zufall, ein zweckloses Ereignis, durch welches in dem thierischen Körper der gesetzmässige Ablauf einer Reihe von Erscheinungen, deren sichtbares Resultat die Geschwulst ist, angeregt wird. Die Pathogenie kann demnach keine andere Aufgabe haben, als jenen Zufall kennen zu lernen and die Gesetze, nach denen die späteren Erscheinungen verlaufen, zu ergründen."[7]

A pure accident is morally neutral; it affixes no blame to the individual. Neither in oncology nor in the study of other diseases do we now take such ready recourse to accident. No appeal to psychosomatic medicine is needed to realize that people rarely fall ill accidentally. Genetic, environmental, and psychological factors—we even speak of accident-prone persons—usually are involved, directly or indirectly. In most illnesses, something or somebody is responsible, from the "sinning" husband who infects his "innocent" wife, to the cigarette smoker who develops cancer of the lung, to the parents who pass on their defect to their children, to the drunken driver who maims a pedestrian, to the government that has failed to provide adequate public health services, and so forth. Responsibility is, of course, a moral issue. All this leads to the conclusion that the pathologist's amoral concept of disease is but part of a more comprehensive concept that includes moral evaluations.

⇥ • ⇤

It is a truism, accepted by most of us, laymen and physicians alike, that disease is an undesirable event. It causes pain, impedes biological and social functions, threatens life, and is a burden to family, friends, and others.[8] In contrast to health, which is assigned a positive value, disease has negative value. Without this broad concept of disease, it would not be possible to distinguish between healthy and morbid, normal and abnormal, and to have a science of pathology. The older Virchow rejected Henle's definition of disease and defined it as "life under *dangerous* circumstances (Leben unter *gefährlichen* Bedingungen)." "Da erwägen wir jedesmal die Beziehungen auf den Organismus im Ganzen, wir untersuchen den Wert, welchen der Vorgang für den Organismus als Ganzes hat; bei der blossen Pathologie kümmern wir uns häufig um das Ganze gar nicht, es ist vollständig ausreichend, wenn wir die Abweichungen [deviations] an den einzelnen Stellen erkennen."[9]

Deviations do not yet constitute disease.[10] Only a teleological concept of the organism as striving to preserve the life of the individual and of the species makes it possible to separate disease from mere statistical abnormality.[11] Disease being an undesirable event, we tend to demand its prevention, treatment, and cure wherever possible and to put blame on whatever or whoever endangers health. Gods, demons, administrators, spouses, parents, landlords, slumdwellers, plant managers, wars, the capitalist system, filth, patients, quacks, physicians, and whatnot were or are held responsible, depending on the kind of disease, the circumstances of its occurrence, our beliefs, our knowledge (pretended or real), the degree of misery, of our wish for improvement, and of our zeal for tracing the culprits. Physicians, together with other health professionals, have been assigned the task of taking care of disease and of the sick and of promoting health.

Seen from this point of view, the study of pathology and other medical sciences, amoral as far as possible, appears as the best way to acquire and promote the knowledge needed in the fight against disease. Detached as a scientist, the physician as healer, counselor, and public health provider shows moral involvement. The medical "scientist" Galen stressed the need for the dispassionate pursuit of truth, including the causes of health and disease. But the hygienist Galen blamed "intemperance, ignorance, or both" for the suffering, helplessness, and consequent indignities brought about by gout, severe arthritis, lithiasis, and indigestion. Such conditions

in a person with a good original constitution were shameful, and "it were better for anyone not an utter coward to choose ten thousand times to die than to endure such a life."[12] Accusing a person's lifestyle as responsible for his diseases has a modern ring, whereas branding as cowardice the preference for life in such conditions over suicide marks the pagan who saw no virtue in bearing patiently the consequences of one's own errors.

William Osler, whose chapter on tuberculosis in his *Principles and Practice of Medicine* was written with the detachment usual in textbooks, at a public meeting in Baltimore violently attacked the mayor, the city council, and the citizens for their neglect to take measures against the disease and in favor of its victims.[13] Few modern physicians will watch with equanimity a diabetic patient disobey their dietetic prescriptions. If a patient refuses a life-saving operation, the physician may even feel in duty bound to appeal to the courts. He may think he is acting wrongly if he allows the patient to die.

Right and wrong actions correspond to what is believed to be good or bad. Thus far, we have taken disease as an unqualified evil that has to be opposed, especially by physicians and public health officials. But even physicians may have to weigh the value of health and disease against other values. The late Esmond Long, a noted phthisiologist, pondered over the doubtful chances of treatment by complete rest if enforced upon the tuberculous composer Chopin. "Of one thing we may be almost sure: had Chopin followed such regimen faithfully over the long period required, and to the exclusion of all else, the world would now be without some of the most treasured of its musical compositions."[14] What is at issue here is which of the two evils is the greater, the personal disease of the composer or the loss of beautiful music. This is but one example out of many more commonplace ones where a similar moral dilemma will arise. It compels us to reconsider the position of disease in the scale of values. Values vary from absolute to relative to nil, and here we have to cite the evaluation of disease ranging from a good (in contradiction to the common assumption) to an absolute, unmitigated evil.

→→ • ←←

The Russian novelist Dostoevsky claimed that in his epileptic aura he had perceived God and felt supreme bliss. It was during such a moment, he maintained, that Mohammed visited paradise. This bliss, Dostoevsky

added, "I would not exchange for all the joys which life can give." Similarly, the epileptic Kirilov in *The Possessed* was willing to give away his whole life for his experience, "because it is worth it," and Prince Myshkin in *The Idiot* defended its value in spite of its morbid origin, of which he was well aware.[15]

For others than himself, Dostoevsky's epilepsy gains value only if it can be said to have contributed positively to the great novels of his later years by opening psychological insights that, as a healthy author, he would not have obtained. The uncertainties of such an assertion even in this inviting case illustrate the problematic nature of the creative power of disease, a theme that has been debated ever since Aristotle (or more likely his pupil Theophrastus) asked why "all men who have become outstanding in philosophy, statesmanship, poetry or the arts are melancholic."[16] The interest was enhanced by the Romantic fascination with disease, particularly tuberculosis, in the early nineteenth century.[17]

Side by side with Dostoevsky's epilepsy, Beethoven's deafness may be cited. Whether it had objective value by giving greater depth to his later compositions has to be decided by musical experts. But the emotional impact of his deafness is beyond doubt. He experienced it as a fate from which there was no escape, which had to be met in a spirit of victorious defiance, a spirit manifested in his fifth symphony, the so-called fate symphony.[18] If we are convinced that his victory over fate outweighed the psychological harm his deafness may have done him, we shall say his disease was good for his soul.

Disease as a boon for the soul is an old idea. *Pathemata—mathemata*, sufferings are lessons, is an old Greek saying.[19] The lessons can take the form of a conversion from a frivolous way of life, as in the case of Francis of Assisi, or of a revelation of truth, particularly religious truth, as for Mary Baker Eddy, the founder of Christian Science. According to her own account, an injury on the ice had brought her near death when her eyes fell upon a biblical passage revealing to her the spiritual essence of man and the illusionary, because material, nature of disease.[20] In severe disease, as in other mortal dangers (e.g., violent storms), it probably is the real or imagined confrontation with death that has the incisive effect.

Judaism has been able to see value in disease, and Christianity has been said to glorify it. Job 2:5–10 relates the story of the righteous man whose loyalty God tests by allowing Satan to smite him with disease. And while

the idea of sickness as divine punishment for sin is very old, the punishment can be borne willingly as atonement and redemption of sin or as a cure of it. "For whom the Lord loveth, he correcteth, even as a father the son in whom he delighteth" (Prov. 3:12). Suffering can purge. "But can suffering actually be very good?" asked Rav Huna, and he answered: "Yes, because through suffering men attain the life in the world to come."[21] It was in the logical extension of such thoughts that lepers were assured that they were already in purgatory.[22]

These ideas of the value of suffering found their supreme example in Jesus, who was believed to have redeemed mankind by suffering for its sins on the cross. Hence, disease becomes a cross to be borne or even to be welcomed by holy men such as Simeon Stylites, who spent many years standing on a column with dire consequences.[23]

It has been said that such ideas have shaped "the Western tradition . . . of the 'ennobling effect of suffering.'"[24] In modern times this notion has been upheld, as far as disease is concerned, on a nonreligious basis by Thomas Mann. To be sure, Mann thought that disease was hostile to man's dignity because it threw him back upon his body. But it also could distance him from nature by spiritualizing him and, as in the case of the tuberculous Schiller and the epileptic Dostoevsky, it could bestow nobility (Adel), a higher form of humanity.[25] In *Der Zauberberg* (*The Magic Mountain*), the hero, Hans Castorp, speaks of the respect due to disease only to be vehemently contradicted by Settembrini, the spokesman for enlightenment, who declares disease a humiliation and the honor paid to it intellectually a profound aberration (*Verirrung*). "Sprechen Sie mir nicht von der 'Vergeistigung,' die durch Krankheit hervorgebracht werden kann, um Gottes Willen, tun Sie es nicht! . . . Ein Mensch, der als Kranker lebt, ist *nur* Körper, das ist das Widermenschliche und Erniedrigende,—er ist in den meisten Fällen nichts Besseres als ein Kadaver."[26]

Mann's work expresses the highest tribute to disease as well as its radical rejection. But on reading the above passages, one may ask oneself whether he was not unduly preoccupied with the potential value of disease for the individual. They may remind us that the period of romantic glamorizing of disease also saw the rise of the industrial revolution, when disease multiplied among the workers in the cities and brought much misery to them. The evaluation of disease is relative to what is believed to be a higher good or a greater evil and depends on the religious, philosophi-

cal, and social orientation of those who judge. Even medical catastrophes have been prized as checking population growth and weeding out the unfit.

Instead of losing ourselves in contemplation of such possibilities, we had better try to integrate what we have said thus far. Characterization as an undesirable evil is essential for the concept of disease, distinguished from other evils by its biological roots, which allow a scientific, amoral study of disease processes. Cultural, economic, and social factors play their role in what is to be considered a disease, so that there is an interplay between the predominating, biological factors and the historically changing ones, neither side having the field entirely to itself. And much as societies are composed of groups and individuals that have particular interests, as human societies they have a common bond. All this resists a radical historization of disease and accounts for the far-reaching agreement in what has counted as sickness, compared to the diversities on which historians like to dwell.

Since by definition disease is something bad, an absolutely good disease would be a contradiction in terms, though disease can be an absolute evil. In between, it can be seen as a relative good, though this does not eliminate the coexistence of suffering, shortened life, and the burdening of others with sorrow, labor, and cost.

In this perspective, the boundary between the "essential nature" of disease as a biological process and any moral attributes attached to it no longer is as sharply drawn as it appeared at first. The recognition of syphilis as a venereal disease determined the personal character of its moral stigmatization, but it also put it in the same class with gonorrhea, thereby for centuries confusing the pathology of these two diseases. Nevertheless, for the pathologist it is not essential how syphilis is contracted, whereas for the public health worker its venereal character is essential because it guides the fight against its spread.

This leads to a final point. We discussed the value of disease in general terms. However, every disease carries its own potential for praise and blame. We do not easily think of dysentery as carrying spiritual value. And while the pain of gout may be good for the soul of an old gourmand, mental disease, at least frank psychosis, is not. Indeed, "moral nosology"—if this name may be permitted—as a systematic study of the stigmata and values that have been connected with individual diseases remains a promising field for the history of medicine.

NOTES

Under the title "The moral concept of disease," a preliminary and rather different version of this essay was presented at the seminar of the Johns Hopkins Institute of the History of Medicine. I thank all those participants who by their questions and comments helped me toward a clearer exposition. The essay integrates as well as elaborates certain ideas with which I previously dealt more or less extensively in the following articles: "Zur Geschichte von 'Moral und Syphilis,'" *Sudhoffs Arch* 19 (1927): 331–48 (English trans. in my book, *The Double Face of Janus and Other Essays in the History of Medicine* [Baltimore: Johns Hopkins Univ. Press, 1977], 472–84); "Medicine and the problem of moral responsibility," *Bull Hist Med* 23 (1949): 1–20 (reprinted in *The Double Face of Janus,* 50–67); "Health and disease," in *Dictionary of the History of Ideas,* ed. Philip P. Wiener, 4 vols. (New York: Scribner, 1973), 2:395–407 (reprinted in *The Double Face of Janus,* 419–40). I have not been able to do justice to the large number of publications that have appeared since then. Besides the articles cited in the notes, I refer for pertinent discussion and literature to the section "Health and Disease" in the *Encyclopedia of Bioethics,* ed. Warren T. Reich, 4 vols. (New York: Free Press, 1978), 2:579–606, and Edmund D. Pellegrino and David C. Thomasma, *A Philosophical Basis of Medical Practice* (New York: Oxford Univ. Press, 1981).

1. *Dorland's Illustrated Medical Dictionary,* 25th ed. (Philadelphia: Saunders, 1974), 1148.

2. Tristram H. Engelhardt Jr., "The disease of masturbation: values and the concept of disease," *Bull Hist Med* 48 (1977): 234–48, showed in detail how the notion of masturbation as a disease "was not value free but structured by the values and expectations of the times" (246). Karl Figlio, "Chlorosis and chronic disease in 19th century Britain: the social constitution of somatic illness in a capitalist society," *Int J Health Services* 8 (1978): 589–617, pointed to the influence of middle-class ideals and prejudices on this alleged disease. How misleading figures for the radical distribution of a disease (e.g., tuberculosis) can be becomes manifest when the incidence among corresponding income groups is compared. Susan Sontag, *Illness as Metaphor* (New York: Farrar Straus and Giroux, 1978), showed the wide influence of psychological, economic, and social beliefs on the comprehension of disease and, vice versa, the use of disease as metaphor.

3. See William F. Bynum, "Chronic alcoholism in the first half of the 19th century," *Bull Hist Med* 42 (1968): 160–85.

4. Benjamin Rush, *An Inquiry into the Effects of Ardent Spirits upon the Human Body and Mind,* 6th ed. (New York: Davis, 1811).

5. The following rough and schematic outline rests on my article on morality and syphilis in *The Double Face of Janus,* 472–84.

6. See Owsei Temkin, *The Falling Sickness: A History of Epilepsy from the Greeks to the Beginnings of Modern Neurology,* 2nd ed. (Baltimore: Johns Hopkins Press, 1971), 383–84. "The disease itself" as an idea enables us to do scientific work as distinct from the historical and sociological approach. Its necessity evinces from the fact that, even to lay bare the biases that made diseases out of masturbation and chlorosis, their status in present medicine must be known. Difficulties arise only when we look upon disease and history as separated in airtight compartments.

7. Rudolf Virchow, "Über die Reform der pathologischen und therapeutischen Anschauungen durch die mikroskopischen Untersuchungen," *Virchows Arch* 1 (1847): 230. For an English translation of the passage, see *The Double Face of Janus,* 62.

8. For more differentiated definitions of disease, see Karl Rothschuh, "Der Krankheitsbegriff (was ist Krankheit?)" *Hippokrates* 43 (1972): 3–17, and Tristram H. Engelhardt Jr., "Ideology and etiology," *J Med Philos* 1 (1976): 656–68.

9. Rudolf Virchow, "Über Akklimatisation," in Karl Sudhoff, *Rudolf Virchow und die deutschen Naturforscherversammlungen* (Leipzig: Akademische Verlagsgesellschaft, 1922), 221.

10. It remains unclear to me whether by deviations Virchow here had in mind minor strayings from a norm of the body as it should be or mere statistical variations, or whether he did not draw a sharp distinction between the two.

11. Cf. Pellegrino and Thomasma, *A Philosophical Basis,* 117. The teleological concept of the organism, not to be confused with vitalistic, prevails among most biologists, though it will be qualified by recourse to Darwinistic explanations.

12. Galen, *De sanitate tuenda* 5.1; *Corpus Medicorum Graecorum* 5.4.2; p. 137, 21–33. The translation quoted is by Robert Montraville Green, *A Translation* of *Galen's Hygiene* (Springfield, Ill.: Charles C. Thomas, 1951), 189.

13. Harvey Cushing, *The Life of Sir William Osler,* 2 vols. (Oxford: Clarendon Press, 1925), 1:570–71.

14. Esmond R. Long, *A History of the Therapy of Tuberculosis and the Case of Frederic Chopin* (Lawrence: Univ. Kansas Press, 1956), 35.

15. For details see Temkin, *The Falling Sickness,* 324, 393, and 394. According to Henri Gastaut, Dostoevsky's ecstatic aura was fictional; see *Epilepsia* 19 (1978): 186–201.

16. Aristotle, *Problems* 30.1 (Loeb Classical Library) 2:155 (W. S. Hett's trans.).

17. See Sontag, *Illness as Metaphor,* 32 f. and passim.

18. English translations of the Heiligenstadt Testament of October 6, 1802, in which he expressed his despair, as well as of the pertinent part of his letter to Wegeler, of November 1801, in which he said "I will seize Fate by the throat," can be found in George R. Marek, *Beethoven: Biography of a Genius* (New York: Funk and Wagnalls, 1969), 325–27, 211.

19. For sources see Klaus Bartels and Ludwig Huber, *Veni, vidi, vici: Geflügelte Worte aus dem Griechischen und Lateinischen* (Zurich: Artemis, 1966), 13.

20. See George Pickering, *Creative Malady* (New York: Oxford University Press, 1974), 193, and Julius Silberger Jr., *Mary Baker Eddy: An Interpretive Biography of the Founder of Christian Science* (Boston: Little, Brown, 1980), 96–109, which suggest that the injury sustained was minor.

21. Quoted from John Bowker, *Problems of Suffering in Religions of the World* (Cambridge: Cambridge Univ. Press, 1970), 35.

22. Guy de Chauliac, *Ars chirurgica* (Venice: Juntas, 1546), fol. 58v. It is also claimed there that "God ... loved Lazarus the Leper more than the others"; see *The Double Face of Janus*, 55–56.

23. Hans Lietzmann (ed.), *Das Leben des heiligen Symeon Stylites: Mit einer deutschen Übersetzung der syrischen Lebensbeschreibung und der Briefe* von Heinrich Hilgenfeld (Leipzig: Hinrich, 1908) (Texte und Untersuchungen zur Geschichte der altchristlichen Literatur. Dritte Reihe, zweiter Band), chaps. 83–91, pp. 129–34.

24. Bowker, *Problems of Suffering*, 15 f.

25. Thomas Mann, "Goethe und Tolstoi," in *Leiden und Größe der Meister* (Frankfurt: Fischer Bücherei, 1957), 35–144; see 55–56 (chap. "Krankheit").

26. Thomas Mann, *Der Zauberberg*, 2 vols. (Frankfurt: Fischer Taschenbuchverlag, 1975), chap. 4 ("Notwendiger Einkauf"), 1:103, 105–6.

The History of Science

CHAPTER 5

Science and Society in the Age
of Copernicus

Those of us who can look back over the past seventy years may well claim to have witnessed a period unique in the history of mankind. If we look for another seventy years equally fraught with momentous events, none perhaps can compete with the age of Copernicus, the seventy years from his birth in 1473 to his death in 1543.

The discovery of a new continent,[1] the first journey around the earth, the rise of Spain to a world monarchy on which the sun never set,[2] the wars that put the Ottoman Turks before the gates of Vienna, that carried foreign armies into Italy, ruined the country, and sacked Rome herself in 1527, the unrest of the peasants and of various factions in the cities, the economic preeminence of Italian and then of German cities (especially Nuremberg and Augsburg), the influx of silver from the New World, which was concomitant with a price revolution, the spread of books through the printing presses, books in the vernacular, in Latin (the language of clerks, philosophers, doctors, and lawyers), and in Greek (the proud possession of the scholar), and finally the Reformation—What other age indeed could compete with ours?

Then, as now, people reacted to events, to those created by themselves as well as those imposed upon them.[3] The increasing importance of siege artillery gave rise to fortifications with bastions from which the defender's cannons could sweep the attacker.[4] Following the foreign invasions, Italian engineers became experts in the planning of fortifications. The firearms themselves also received due attention. In a book of 1537, Nic-

Originally published in *The Nature of Scientific Discovery*, edited by Owen Gingerich (Washington, D.C.: Smithsonian Institution Press, 1975).

colò Tartaglia tells of a bombardier's question of how to aim a cannon "for its farthest shot." Tartaglia, "by physical and geometrical reasoning," proved that an elevation of 45 degrees above the horizon would be optimal.[5] The problem led Tartaglia to theoretical considerations about natural and violent movements, the former represented by freely falling bodies, the latter by projectiles,[6] to the invention of various explosives and other things. He "was going to give rules for the art of the bombardier and bring this to the greatest possible detail" when the thought struck him that he was engaging in a matter "cruel and deserving of no small punishment by God." He stopped, destroyed what he had done, and was full of regrets. But when the Turks threatened Italy, Tartaglia reconsidered: "Seeing that the wolf is anxious to ravage our flock ... it no longer appears permissible to me at present to keep these things hidden."[7]

The voyages of exploration naturally led to charting and map making, and these in turn were connected with astronomical observations and astronomical instruments. The close connection between geography, which had to define the latitude and longitude of localities, and astronomy is seen in the term *cosmography*, which could refer to the terrestrial world as well as the cosmic.

Another practical aspect of astronomy was calendar making, and the German calendar by Regiomontanus may be taken as an example of what interested its users. It begins with 1475, the year before Regiomontanus's death, and takes Nuremberg for its geographic base. It allows calculation for many years to come and for many towns and countries. Latin dates corresponding to the days of the month, the dates of the holidays, the times of the new moon and the full moon, the positions of sun and moon, all can be calculated. The large space given to eclipses of sun and moon is surprising. The location of the moon within the signs of the zodiac was important because these signs were credited with qualities, and they were correlated to the various parts of the body. The reader was informed whether the lunar position was, or was not, favorable for bleeding. For instance: "As the doctors tell us, the Ram (Aries) is hot, dry, and fiery. Man's head is associated with it, and it [i.e., the time, not the head] is convenient for bleeding."[8] Regiomontanus also informs the reader that young people are usually bled at the time of the waxing moon, old people at the time of the waning moon.

Now Regiomontanus, besides being one of the greatest astronomers

FIGURE I. Methods of descending into mines. From Georg Agricola, *De re metallica* (Basel, 1556).

and mathematicians of his day, also was a very learned man. His calendar appeared in Latin and German versions.[9] The calendar proved very popular, though the demands it made on the interests and the intelligence of the reader must not be underrated; they are quite comparable with the trust our government puts in the citizen's ability to fill out his tax forms correctly.

The importance of mining was reflected in the work of the physician

and classical scholar Georg Agricola.[10] His first book on mining was a dialogue between an employee at the mines of Joachimsthal, now Jachymov, in Bohemia, and two doctors of medicine. The employee, Bermannus, whose name gave the book its title, is the expert who demonstrates various minerals to the others. Together they discuss their properties and their relationship to the minerals described by Dioscorides, Vitruvius, Pliny, and Galen. Besides being interested in mathematics, geometry, poetry, and astronomy, Bermannus reveals himself as a Latin scholar of no mean rank—but he does not know Greek and is not at home in the writings of the Arabs.[11] The book is a preliminary study to Agricola's more famous treatise; it aims at arousing interest in minerals and their medicinal properties. The ancients knew many of them—but now ignorance prevails; neither apothecaries nor physicians have any idea of what the substances they read about are like. Moreover, German mines contain things of which the Greeks were ignorant and that should be made known.[12]

Bermannus is a realistic book. It names persons who owed their wealth to the mines; it mentions the yield in figures, it mentions new machines; it touches on the history of the mining towns: twelve years before, there had been only one house in Joachimsthal next to an old dilapidated pit—now the town was buzzing with miners and their activities.[13] The same realistic attitude prevails toward the minerals: their identification is the main point. The books of the ancients and of the Arabs are the existing reference works giving basic information. The Arabs knew substances unknown to the Greeks, and now there are others to be added. Agricola complains about the repetitiousness in quoting the ancients; a good compendium from their best authors would suffice; only new things are worth saying about medicine.[14] Agricola's wish to see medicine brought back to the high standard of antiquity means a return to a progressive study of things as they are, without indulgence in subtleties.

When we think of the period between 1473 and 1543, our first association probably will be with beautiful pictures, statues, and buildings rather than with cannons, calendars, mining, and other technical matters.[15] Indeed, in the artist, so we are told, we meet the predecessor of the modern scientist.[16] The artist wished to show things as they "really" were, and he was thus led to ask what things were really like and how they could be represented so as to be seen as they are.[17] In particular, the artist studied the human body, and he studied proportions and perspective, both of

which rest on mathematics. And since the artist of that age also was architect and engineer, he appears as the early embodiment of the triad: knowledge of nature, mathematics, and technology. We have only to think of Leonardo da Vinci, who left anatomical drawings unequaled in their truthfulness by anything anatomists of his age had to offer, who recommended himself as an army engineer to the Italian princes, and whose imaginative power ranged far into the fantastic. We remember that, beyond the Alps and somewhat later, Albrecht Dürer, born two years before Copernicus, not only worked on fortifications but also wrote on human proportions and on mathematics. Recent studies on the seventh book of the *Trattato di architectura* by Francesco di Giorgio Martini have revealed drawings of various mechanical devices for civilian use, especially mills, the main inanimate power machine, drawings from which later authors borrowed freely.[18] Most artists of the time lacked formal classical education.[19] They learned their craft from their masters but, conscious of their knowledge, many of them "fought the great battle for the recognition of painting as one of the 'liberal' arts," to quote the late Erwin Panofsky.[20]

All this agrees well with the discovery by Andreas Vesalius that Galen's anatomy was based on animal dissections and that human anatomy had to be based on a careful study of human cadavers. By a curious coincidence, Copernicus's *De revolutionibus*, offering the new view of the macrocosm, and Vesalius's *Fabrica*, containing Galen's refutation and many new anatomical discoveries about the microcosm, man's body, appeared in the same year. Vesalius demonstrated that anatomical truth about man had to be derived from autopsy (i.e., inspection of human bodies), as Agricola was teaching with regard to mineralogy and Fuchs and others with regard to botany.[21] Text and illustrations of the *Fabrica* view the body realistically in contrast to its former schematic representations. The printing press assured that all readers could see the same illustrations, and Vesalius himself emphasized the social ties of anatomy. Contempt for manual work in medicine (i.e., surgery and anatomy), he thought, was responsible for their decline.[22] He, too, hoped for their restoration to the glory of the ancients, and even more than Agricola he returned to antiquity via a devastating critique of ancient authority, in this case, Galen.

What we have done thus far is to choose a few works we could easily call "scientific" because they were mainly occupied with natural facts and mathematics and relate them to the needs of the time. This relationship

FIGURE 2. Artists preparing pictures of plants. From Leonhart Fuchs,
De historia stirpium commentarii (Basel, 1542).

was technological in a very broad sense; directly or indirectly, these books
intended to serve political, economic, medical, and esthetic needs of man.
To us, the technological aspect seems characteristic of science. It pleases
us when we think of the good science can do, and it frightens us when
we think of the evils.

In the age of Copernicus, technology was also associated with work
and ideas that do not fit into our picture of science and which, for brevity's
sake, we shall call *magic*. Magic was woven of many threads: the old belief
in the occult forces of nature and the equally old belief in the influence
of the stars; the old fear of witches and devils, powerfully documented
in the *Malleus maleficarum* of 1485 (the *Witches' Hammer*), which made discovery of witches a fine art; the study of the Hermetic Books of late antiquity, which Marsilio Ficino introduced into Neo-Platonic philosophy;
the study of the Cabala; and a whole body of popular beliefs taken seriously by intellectual leaders of the time. When about twenty-three years
old, the arch magician Agrippa of Nettesheim wrote a book *De occulta medicina* (*On Occult Medicine*), which he published later, in 1533. Agrippa acknowledged himself to be a magus (a magician), one of those who
thought it possible to ascend from the world of elements to that of the

stars and hence to the intellectual world and to God himself. The magician learned virtues of things, which empowered him to prognosticate and to perform astonishing deeds. Agrippa denied being a sorcerer; he freely admitted that the study of magic taught forbidden things. "But those things which are for the profit of men—for the turning away of evil events, for the destroying of sorceries, for the curing of diseases, for the exterminating of phantasms, for the preserving of life, honor, or fortune—may be done without offense to God or injury to religion. Who will not deem them just as profitable as necessary?"[23] The magician's confidence in the ability to operate with the magic forces that build the universe led a modern philosopher to say that "magic . . . was a set of rules for gaining power over the world, and that was also Bacon's program and has remained the program of the applied scientists, engineers, physicians, and advertising men."[24]

Nevertheless, the temptation may be great to dismiss all magic, both black and white, and all divination including astrology, and all belief in demons and witches by lumping them together as pseudoscience or superstitions. We can do this for our own times, but we cannot do it for the age of Copernicus.[25] Where would we draw the line? Bermannus speaks of the demons in the mines, some harmless and some evil.[26] Regiomontanus relates the astrological significance of the signs of the zodiac. If we eliminate these matters as "superstitions," the remainder may be a pure distillate of science as we understand it. But our treatment of science and society would become unreal. We cannot disregard Paracelsus, the most colorful medical figure of the age of Copernicus, simply because he, too, was a magus; nor can we divide his work and say these books and chapters we accept, those we disregard. The original Doctor Faustus lived in the age of Copernicus, and his spirit refuses to be exorcised.

What then was meant by science? Linguistically, the answer is simple. *Scientia* meant knowledge—more particularly, knowledge methodically acquired and taught—and there were many *scientiae,* many kinds of knowledge. But what we call *science* did not exist as a separate discipline; its use for the time is essentially anachronistic, as has been pointed out by others before me.[27]

There existed the mathematical disciplines, including astronomy; there existed natural philosophy that ranged from a knowledge of the elements to that of the soul as a physiological and psychological entity; and there

existed a metaphysical realm from which physics was not clearly separated. There were physicians who, as the name indicates, were students of nature (*phusis*), there were mathematicians and astronomers or astrologers, there were teachers of philosophy, but "scientists" as a group or profession did not yet exist. In recent years much has been written about the preoccupation with methodology, especially in Padua, where arts and medicine were closely associated and where the so-called scientific method, later applied by Galileo, is said to have been advanced theoretically. Professor Hall has already analyzed these discussions. Whatever their later influence may have been, in the age of Copernicus these debates hardly led to scientific discoveries, though they found an echo in Copernicus himself.[28]

Discovery is inbuilt within the very notion of modern science, where knowledge is not allowed to remain static. Experiments are devised so as to give new answers to what is hypothetical or to establish facts that will verify or falsify theories. A theory, be it ever so beautiful, is thought unsatisfactory if it is without experimental or observational consequences. But in the age of Copernicus, knowledge, including knowledge of nature, was still taught as a traditional body, largely on the basis of authoritative texts. In the *Margarita philosophica* of Reisch, an encyclopedia in Latin (i.e., for the educated), the subdivisions of natural philosophy consist mainly in the titles of Aristotelian books.[29]

For those who could not read, knowledge of nature was the practical knowledge required for tilling the land, raising cattle, and practicing one's craft, spiced by homely wisdom, the preacher's word, pictures in the church, and biblical tales. Those who could lay claim to some education participated through popular works and to varying degrees in the teaching of the schools.[30]

In the schools, the world was accepted as created by God yet understandable as a coherent system based on metaphysical notions of matter and form, on four causes (among which the final cause loomed large), on the four elements of air, earth, fire, and water, themselves combinations of the four qualities of hot, cold, dry, and wet.[31] Not only was the earth, and with it man, in the center of this universe, but what happened was understandable because much of it was expressed in terms of human experience. To say with Aristotle that things were heavy because they tended downward to their natural place and light because their tendency was up-

FIGURE 3. Dance of the Dead. From Hartmann Schedel, *Liber chronicarum* (Nuremberg, 1493). Compare these skeletons with the drawing from Vesalius.

ward made such phenomena more human than a mathematical law of gravity, which does not say what gravity is.[32]

In interpreting prescribed texts, professors could point out contradictions and could deviate from their authorities. Thus, some medieval schoolmen had changed Aristotle's doctrine of violent motion and had found new formulations for accelerated motion.[33] New interpretations and new observations might meet with resistance, but they were not barred; nor, on the other hand, were they expected, let alone demanded.

Works on education allow some insight into the role allotted to mathematics and natural philosophy among the cultivated class. Children, said

Leon Battista Alberti, must learn to read and write. "Afterwards they should learn the abacus and, at the same time, they may also look at geometry as far as it may be useful: these two are sciences that are proper and pleasing for children's minds and no little useful in every circumstance and age."[34] That is little enough, even for children, coming as it does from an architect and theoretician of the arts, especially if compared with Alberti's enthusiastic praise of the study of literature, above all Latin literature. Alberti died a year before Copernicus was born; his thoughts may be considered influential at the time of Copernicus's youth. Juan Luis Vives, on the other hand, published his great pedagogic work *De tradendis disciplinis* in 1531. According to him, the chief aim of education was religious, moral, and civil. Within this framework, the practical uses of the astronomical sciences and of natural philosophy were stressed, and Vives was particularly interested in such matters as medicinal plants, navigation, architecture, and all crafts. They need not, however, be taught in school; rather, the taste to acquire such knowledge should be cultivated.[35] But Vives also warned that knowledge gained of nature "can only be reckoned as probable and not assumed as absolutely true." Men, like Pliny and Aristotle, who insist on sensual or logically indisputable causes for everything "become incredulous of the discoveries of others, and unbelieving in matters of religion."[36]

Vives belonged to the circle of Sir Thomas More and Erasmus of Rotterdam. In More's *Utopia*, "science" does not play any conspicuous role. Erasmus, on the other hand, in a short letter that was prefaced to Agricola's *Bermannus*, allows a glimpse of his attitude. The book pleased him, and the veins of silver and gold almost induced in him a desire for such things. "If only," he exclaims, "our minds carried us heavenward with the same zeal with which we search the earth." There is nothing wrong with the mining industry, but "only the vein of sacred books can truly enrich man."[37]

Some three hundred years later, Dr. Arnold, the headmaster of Rugby, wrote to a friend:

> Rather than have physical science the principal thing in my son's mind, I would gladly have him think that the sun went round the earth, and that the stars were so many spangles set in the bright blue firmament. Surely the one thing needed for a Christian and an Englishman to study is Christian and moral and political philosophy.[38]

In the days of Dr. Arnold, science did exist, and it posed a serious threat to fundamentalist beliefs and to the primacy of a classical and moral education. In the days of Vives and Erasmus, no such threat existed—the natural knowledge of the time was itself full of miraculous stories. The various branches of the pseudosciences were controversial, and there were those who rejected the casting of horoscopes and any form of black magic, but there was near unanimity in the admission of white magic, the knowledge of and command over the occult forces of Nature. How could it be otherwise, when all religions firmly asserted the historical reality of miracles that were not verifiable? Their acceptance presupposed a habit of mind that did not insist on verification of alleged facts of experience, if vouched for by the authority of tradition or of a venerable name.

Be that as it may, tradition and authority, we shall concede, had their place in theology and the law, both of which relied on the authorities of the past: the Bible, the fathers and councils of the church, the code of Justinian, and the canon law. Here, and perhaps in metaphysics, was a place for interpretation by argument. But it ought to have been clear, we may think, that such methods and beliefs were futile in the explanation of Nature, which had to be discovered rather than interpreted.

In our impatience with authority and tradition in science, we may well ask what happened when they were seriously challenged. For such challenge and pressure for new discoveries were not altogether lacking, as our discussion of the great magi suggested and for which Paracelsus is the prime example. Paracelsus condemned the medicine and natural philosophy of the schools; he appealed to experience in the light of nature; to the Aristotelian elements he opposed the chemical principles of salt, mercury, and sulfur; and he urged the finding of cures for allegedly incurable diseases. His declared motives for asking for the new were religious and social. God had not created physicians to let people die. It was their duty to find cures, rather than to grow rich and to parade the insignia of their doctoral rank. Without trust in God, the search would be futile.[39]

Paracelsus met with violent opposition, which, toward the end of his life (he died in 1541), drove him to write his *Seven Defensiones*. Of these, characteristically enough, three are defenses of his discovery of a *new* medicine, of *new* diseases, and of *new* recipes. Much of the opposition to him was shortsighted; but if, for a moment, we imagine that the opposition had failed, the immediate result could hardly be envisaged as anything but

chaos. Paracelsus's writings are very hard to understand, his principles of research lack methodological clarity, and the curative power of his remedies is far from certain. To be sure, authority covered much sham, but it also was intended to protect against unscrupulous empiricism and experimentation. In a popular German therapeutic work of 1524, the author recommended the authority of Hippocrates with these words:

> For many years the art of medicine lay neglected until holy Hippocrates . . . saw the light of the world. And rightly is he called "holy," for God Himself was without doubt in him, so that he gave the human race such fruitful teaching, and none can say that this same Hippocrates erred in any single thing in medicine. He was a pious, virtuous man, not idle, and always wrote for man's well-being.[40]

Generally speaking, experience needed authority; it could not be relied upon outside of a few fields like anatomy and botany. For the rest, moderate antagonism to authority and tradition did exist independently of Paracelsus and before the end of the age of Copernicus. For example, Girolamo Fracastoro wrote a short treatise in which, very apologetically, he dared to take issue with Galen.[41]

Quite possibly Fracastoro and Copernicus were fellow students in Padua when Copernicus studied medicine there. They had more in common than medicine, for Fracastoro also wrote on astronomy. The versatility of the learned corresponded to the versatility of those who expressed themselves in drawing, namely, architects, painters, and sculptors;[42] it corresponded to a lack of specialization that united smiths or gunfounders and clockmakers.[43] Book learning was not an adequate method for experimental science, but it allowed familiarity with many branches of study. Many outstanding philosophers, Marsilio Ficino, Pomponazzi, Achillini, and Zimara, were also physicians, or at least trained in medicine. Others, like Cardanus and the older Scaliger, are hard to pigeonhole, and Copernicus himself held the degree of doctor of canon law. Versatility, of which Fracastoro was an outstanding example, was as much a matter of lacking need for specialization as a manifestation of the urge to universality.

Today, the fame of Fracastoro rests on his study of contagious diseases and on his poem *Syphilis sive De morbo Gallico* because it coined the name syphilis for the disease then known to a majority as the French pox and as the *mal de Naples* to the French. In his own days, this poem was highly es-

teemed for the beauty of its Latin hexameters. Fracastoro also wrote on sympathy and antipathy between things, on intellectual comprehension (which he considered part of natural philosophy), and on the soul.

Fracastoro's astronomical treatise, which preceded Copernicus's *De rev-olutionibus* by five years, may here be cited not for his theory of homocentric cycles but for a sentiment about the usefulness of astronomical study. After stating the many practical uses of astronomy, Fracastoro wrote:

> Besides, this contemplation seems in no small degree conducive to putting our minds at rest, first, because we never recognize our own and our life's insignificance as much as when we admire the greatness and perpetuity of their immortal bodies, also because lifting our mind from these terrestrial affairs we become accustomed to those that are divine. From here first came knowledge of the gods, love for them and veneration, and admiration of eternity.[44]

These thoughts were not altogether original. The Roman philosopher Seneca had voiced similar sentiments some fifteen hundred years before.[45] The earth might be in the center of the world, but together with man it was believed to be of inferior material and insignificant if compared with the divine majesty of the stars. Copernicus moved man and earth out of the center, but he made them the equals of the other heavenly bodies.

How man's image of himself changed with such different cosmological views is an intriguing question. Which of the two was more likely to make him feel proud or humble? Whatever the answer may be, to us the whole subject seems to lie outside scientific astronomy. The effects of science on man are extremely important, but they are only effects; they do not belong to the science itself. Science pursues truth in its own way; its verified results are valid regardless of human feelings, social application, moral and esthetic values, and religious convictions.

Such a detachment may have been in the minds of a few individuals. Here we are not concerned with individuals in social isolation but in their interactions. In such a social sense, I believe, this kind of detachment did not yet exist. Even if the astronomer, the anatomist, or the philosopher was allowed to pursue his research in his own way, it was on condition that in the end any conflict with religion was resolved in favor of the latter.[46] When Osiander, in his preface to the reader of Copernicus's work, claimed that divine revelation alone contained the truth, which neither the

pragmatic hypotheses of astronomers nor the probabilistic speculations of philosophers could offer, he misrepresented Copernicus.[47] As Professor Hall told us in the preceding article, Copernicus's achievement to no small degree consisted in upholding the reality of the heliocentric system.[48] But calling this an achievement underlines the exceptional nature of such independence in controversial matters. Moreover, though yielding to religious truth was the most obvious obstacle to the detachment of science from nonscientific assumptions and values, it was not the only one, and here Copernicus himself can be quoted.

He, too, refers to the smallness of the earth as compared with the quasi-infinite size of the universe, and he uses this argument against the view that this vast world, rather than the small earth, revolved in twenty-four hours. Immobility is held to be more noble and more divine than change and instability, which therefore is more becoming to the earth,[49] and he praises the central position of the sun with words that appeal to old associations linking the sun with power and dignity.

> In this most beautiful temple, who would care to put this lamp [the sun] in any other or better place than whence it can illuminate the whole at the same time? Thus some people not inaptly call [it] the lamp of the world, others [its] mind, others [its] governor. Trimegistus [calls it] the visible god, Sophocles's Electra [refers to it] as seeing everything. Indeed, residing on a royal throne, as it were, the sun rules the revolving family of the stars.[50]

The fact that we can easily disregard such passages does not allow us to think of them as irrelevant for the conviction of Copernicus that the heliocentric world was real and no mere hypothesis. Similarly, the philosophical postures that Vesalius (or his artist) gave to the skeleton and the landscape within which he placed his musclemen hint at the feelings and thoughts associated with anatomy. In the popular dance of death, the skeleton appears as the symbol of death. Anatomically much more exact, the symbolism is still there.

To us, the landscape seems irrelevant for anatomy; the beauty of an anatomical illustration lies in its exactness and clearness. Yet for Vesalius's artist this functional beauty, if I may say so, does not suffice; even a flayed human figure has to be given a place in the world, rather than on a

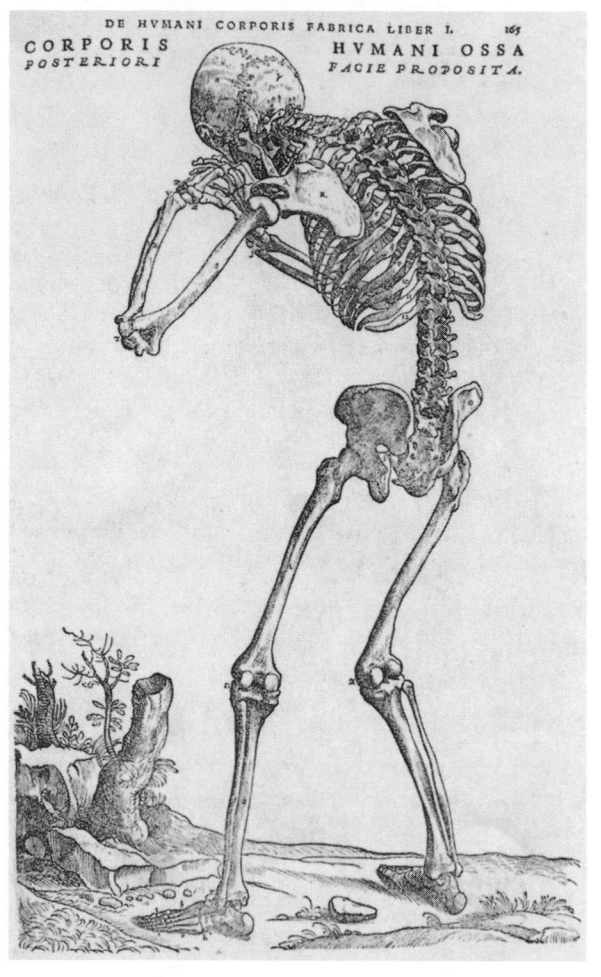

DE HVMANI CORPORIS FABRICA LIBER I.   165
CORPORIS
POSTERIORI
HVMANI OSSA
FACIE PROPOSITA.

FIGURE 4. Human skeleton. From Andreas Vesalius, *De humani corporis fabrica* (Basel, 1543). This great work was published in the same year as Copernicus's *De revolutionibus.*

bare page. This distinguishes the anatomical picture from the schematic drawings in the *Fabrica,* and the incorporation of such pictures is no less remarkable than the writing on syphilis in beautiful verse.

As a last example, a chapter can be cited from the *Narratio prima* of 1540, in which Rheticus gave his preliminary account of Copernicus's astronomy.[51] The chapter is entitled "The Kingdoms of the World Change with

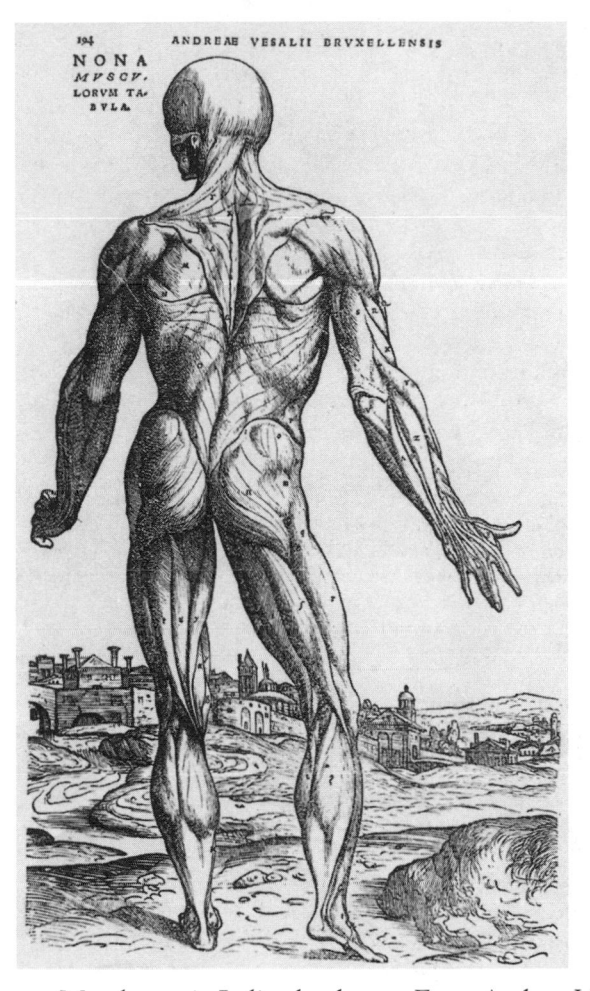

FIGURE 5. Muscleman in Italian landscape. From Andreas Vesalius,
De humani corporis fabrica (Basel, 1543).

the Motion of the Eccentric," the eccentric being the device by which the
varying distances of the earth from the sun were explained.[52] In this chapter, Rheticus establishes a relationship between eccentricity and Rome's
becoming a monarchy, its decline "as though aging," and its fall, the es-
tablishment of Islam, and the extension of the latter's realm (i.e., by the
Turkish empire).

> In our time it is at its pinnacle from which equally swiftly, God willing, it
> will fall with a mighty crash. We look forward to the coming of our Lord

Jesus Christ when the center of the eccentric reaches the other boundary of mean value, for it was in that position at the creation of the world. This calculation does not differ much from the saying of Elijah, who prophesied under divine inspiration that the world would endure only 6,000 years, during which time nearly two revolutions are completed.[53]

Such connections of human history and heavenly events were common enough, and they usually went back to the book of Daniel and the Apocalypse of John. The idea of heaven's having an age was forcefully expressed by Paracelsus: "Now the sky too was a child, it too had its beginning and is predestined to its end, like man, and death is in it and around it."[54]

Are such ideas to be put together with astrology, which exercised so profound an influence on high and low, on rich and poor? Whatever the answer may be, I wish to substitute a more general thought. To the generation that believed in the Bible, God had created the whole world, heavens, earth, plants, animals, and man within six days. Man and the sky have a common creation; should they not also have a common history from creation to the day of judgment?[55] Once such a notion is accepted, parallels between macrocosm and microcosm, between heavenly events and human ones, are not intrinsically fantastic. Constellations in the sky may take on the nature of portents and thus allow prognostications, even if causative power is denied to the stars. Such basic convictions rooted in a religion common to all can be much stronger than speculative theories. They make it difficult to believe that stars and men go their separate ways, that the world of nature is without meaning, without beauty, and without justice.[56]

We have spoken about detaching nature from human values; we have not yet spoken about the psychological detachment that allows man to study the world free from passion. In the age of Copernicus, people by and large more easily gave way to their impulses of anger, aggressiveness, and immediate joy than we do, who are compelled to exert constant control over ourselves and rational foresight.[57] Did those who studied nature and the behavior of man stand out by their dispassionate character?

There is the story of Leonardo da Vinci and the centenarian in the Hospital of Santa Maria Nuova in Florence.

This old man [Leonardo relates], a few hours before his death, told me that he was conscious of no bodily failure other than feebleness. And thus

sitting on a bed . . . without any worrying movement or sign, he passed from this life. And I made an anatomy to see the cause of a death so sweet.[58]

The story has more than anecdotal value when we realize that anatomists who stole bodies from the gallows might have to dissect them clandestinely under conditions that would make their modern colleagues shudder.[59]

The other example concerns Guicciardini, the great Florentine historian and statesman and friend of Macchiavelli. In his *Ricordi,* Guicciardini remarks that people are quite wrong who believe that military victory depends on the justice or injustice of a cause. The belief in the justice of one's cause can contribute to victory because it makes people bold and obstinate. "Hence having a just cause may indirectly help, but it is false to say that it does so directly."[60]

So much these two examples show: There were men in the age of Copernicus who could view the life and the behavior of man dispassionately, even to a frightening degree, as demonstrated by Machiavelli. It may be hard for us to realize that the desire to see things as they are and yet to see them in a framework in which science and pseudoscience, nature and values are not clearly separated can coexist. This difficulty of ours, I think, lays bare some of *our* assumptions about science and, implicitly, about the significance of scientific discovery.

We are celebrating the 500th anniversary of the birth of Copernicus (i.e., we are celebrating the man and not only his work). Therefore, it seemed appropriate to me to write of his age in the narrow limitation of his birth and death. To bring this age into clearer focus, I have, as far as possible, refrained from using such broader concepts as renaissance and humanism, whose importance is beyond doubt but which would blur the limits. In consequence, I have been able to offer no more than an impression, which undoubtedly could be supplemented, perhaps even supplanted, by different impressions based on a different and wider selection of examples and witnesses. The impression I offered was not that of a neat historical entity generated by the preceding age and itself generating the next. Copernicus's book was a product of his own age, but its fate was determined by old and new circumstances and by people of whom some had survived him, others were young when he died, and still others were born after him.

How they dealt with Copernicus's doctrine and why they did so is their own story.[61] By narrowing our view, we may, however, have come a little closer to "the untidiness of human affairs,"[62] which marks reality and which ruled over the relationship of science and society then, as it does now.

NOTES

1. On the interrelationship of the discovery of America and European intellectual and economic life, see J. H. Elliott, *The Old World and the New, 1492–1650* (Cambridge, 1970).

2. The surprising rise of Spain was studied in some detail by J. H. Elliott, *Imperial Spain, 1469–1716* (New York, 1963).

3. Norbert Elias, *Über den Prozess der Zivilisation: Soziogenetische und psychogenetische Untersuchungen*, 2nd ed. (Bonn, 1969), investigated the development of behavior within the political and social transformations of Europe since the Middle Ages.

4. Horst De la Croix, "The literature on fortification in Renaissance Italy," *Technol Culture* 4 (1963): 31 (with regard to Italy): "Not only military men, but humanistic scholars, architects, and artists began to devote their time and energy to an intense study of the problems raised by the apparently irresistible power of the cannon."

5. Niccolò Tartaglia, *Nova Scientia*, quoted from Stillman Drake and I. E. Drabkin, *Mechanics in Sixteenth-Century Italy: Selections from Tartaglia, Benedetti, Guido Ubaldo, and Galileo* (Madison, 1969), 63–64.

6. Ibid., 72 ff.

7. Ibid., 68–69.

8. *Der deutsche Kalender des Johannes Regiomontan, Nürnberg, um 1474*, with an introduction by Ernst Zinner (*Veröffentlichungen der Gesellschaft für Typenkunde des XV. Jahrhunderts, "Wiegendruckgesellschaft,"* series B, vol 1) (Leipzig, 1937), f. 24ʳ.

9. As Zinner (ibid., 9) pointed out, in writing the German edition Regiomontanus had to cope with a language as yet quite unadapted to mathematical and astronomical discourse.

10. For Agricola and his works, see now *Georgius Agricola: Ausgewählte Werke* (Berlin, 1955–). In particular, Vol. 1: Helmut Wilsdorf *Georg Agricola und seine Zeit*, is important for the biography of Agricola, his forerunners in mineralogy and mining, and the conditions of his time. Vol. 2: *Bermannus oder über den Bergbau: Ein Dialog*, contains a German translation of the work by Wilsdorf. For the Latin text I have used the edition in *De re metallica* (and other works) (Basel, 1657), 679–701.

11. *Bermannus*, 70 (German) and 683a (Latin).

12. Ibid.

13. Ibid., 77 (German) and 684b (Latin). According to Wilsdorf, *Georg Agri-*

cola und seine Zeit, 163–64, there were about 900 pits and tunnels in Joachimsthal in Agricola's time; by 1579, however, the number of tunnels had fallen to 13.

14. *Bermannus*, 107, 110, 112 (German) and 690b, 691a, 691b (Latin).

15. For a general survey of technology in our period, see A. Rupert Hall, "Early modern technology, to 1600," in *Technology in Western Civilization*, ed. Melvin Kranzberg and Carroll W. Pursell Jr. (New York, 1967), 1:79–103.

16. Leonardo Olschki, *Geschichte der neusprachlichen wissenschaftlichen Literatur* (Leipzig, 1919–22; Halle, 1927), is fundamental for the study of the artist's role in the Renaissance. See also Erwin Panofsky, *The Codex Huygens and Leonardo da Vinci's Art Theory* (London, 1940), 91.

17. Erwin Panofsky, *The Life and Art of Albrecht Dürer* (Princeton, 1955), 243: "The Renaissance . . . established and unanimously accepted what seems to be the most obvious, and actually is the most problematic dogma of aesthetic theory: the dogma that the work of art is the direct and faithful representation of the natural object."

18. Ladislao Reti, "Francesco di Giorgio Martini's treatise on engineering and its plagiarists," *Technol Culture* 4 (1963): 287–98.

19. For details see Olschki, *Geschichte der neusprachlichen*, 30–46.

20. Panofsky, *Life and Art of Dürer*, 270.

21. The methodological importance of Vesalius lies in his emphasis on learning human anatomy from the dissection of human bodies; see Owsei Temkin, *Galenism: Rise and Decline of a Medical Philosophy* (Ithaca, N.Y., 1973). Characteristic for the realistic attitude of the botanists is a picture in Leonhart Fuchs's *De historia stirpium commentarii insignes* (Basel, 1542), showing two artists at work.

22. See the English translation of the preface to the *Fabrica* by C. D. O'Malley, *Andreas Vesalius of Brussels, 1514–1564* (Berkeley and Los Angeles, 1964), 317–24.

23. Henricus Cornelius Agrippa ab Nettesheym, *De occulta philosophia libri tres* (n.p., 1533), "Ad lectorem," fol. aaii^r. The English translation is quoted (with modification) from *Three Books of Occult Philosophy or Magic by Henry Cornelius Agrippa, Book One: Natural Magic*, ed. Willis F. Whitehead (1892) (reprinted New York, 1971), 26. For the relationship of Agrippa's magic work to his religious endeavors, see Charles G. Nauert Jr., *Agrippa and the Crisis of Renaissance Thought* (Urbana, 1965) [*Illinois Studies in the Social Sciences*, vol. 55]. For Agrippa's place in the Hermetic tradition, see Frances A. Yates, *Giordano Bruno and the Hermetic Tradition* (Chicago, 1964). Wayne Shumaker, *The Occult Sciences in the Renaissance: A Study in Intellectual Patterns* (Berkeley and Los Angeles, 1972), devotes a lengthy section (134–56) to the analysis of Agrippa's work. Shumaker's book is the most comprehensive recent discussion of all forms of "occult sciences" in the Renaissance.

24. George Boas, "Philosophies of science in Florentine Platonism" in *Art,*

Science, and History in the Renaissance, ed. Charles Singleton (Baltimore, 1967), 241. See also Frances A. Yates, "The Hermetic tradition in Renaissance science," ibid., 255.

25. Cf. Hugh Kearney, *Science and Change, 1500–1700* (New York, 1971) (World University Library), 37 ff. and 96 ff., and W. P. D. Wightman, *Science in a Renaissance Society* (London, 1972), chap. 11.

26. *Bermannus,* 88 (German) and 686b (Latin).

27. In "Towards a new history of the new science," *Times Literary Supplement,* 15 September 1972, 1058, the anonymous reviewer, noting the absence, in 1556, of Copernicus's *De revolutionibus* from any Oxford college library, remarked on "the caution which it is necessary to use when accepting the word 'science' in an anachronistic sense." See also Wightman, *Science in a Renaissance Society,* 19.

28. For the debate on "method," see John Herman Randall Jr., *The School of Padua and the Emergence of Modern Science* (Padua, 1961), and Neal W. Gilbert, *Renaissance Concepts of Method* (New York, 1960). Copernicus himself, in the preface to Pope Paul III, *De revolutionibus orbium coelestium,* bk. 6 (Nuremberg, 1543), fol. iii^r, wrote of earlier astronomers: "Itaque in processu demonstrationis, quam μέθοδον vocant, vel praeteriisse aliquid necessariorum, vel alienum quid, et ad rem minime pertinens, admisisse inveniuntur. Id quod illis minime accidisset, si certa principia sequuti essent. Nam si assumptae illorum hypotheses non essent fallaces, omnia quae ex illis sequuntur, verificarentur proculdubio." [Hence in the course of the demonstration, which they call "method," they are found either to have omitted something indispensible or to have introduced something extraneous and wholly irrelevant. This would not have happened if they had followed established principles. For if the hypotheses they assumed were not false, everything that followed would have been verified beyond any doubt.] As Edward Grant, "Late Medieval Thought, Copernicus, and the Scientific Revolution," *J History Ideas* 23 (1962): 213, has shown, Copernicus used the argument to demand of hypotheses that they be true, but it did not lead him to his own discovery.

29. *Margarita philosophica nova* (Strasbourg, 1515), fol. A iv^r: "Philosophiae particio." Theoretical philosophy has two parts: *realis* and *rationalis.* The former is subdivided into *Metaphysicam, Mathematicam* (consisting of the traditional quadrivium, i.e., arithmetic, geometry, music, and astronomy), and "Physicam sive naturalem sub qua et medicina theorica continetur et traditur in libris Phisicorum, De caelo et mundo, De generatione et corruptione." *Philosophia rationalis* is represented by the *trivium* (grammar, rhetoric, and logic). Medicine as such (i.e., without what we would call the basic sciences) belongs to the mechanical arts (*Philosophia factiva*) together with agriculture, navigation, etc.

30. Karl Sudhoff, *Deutsche medizinische Inkunabeln* (Leipzig, 1908) [*Studien zur Geschichte der Medizin,* fasc. 2–3], offers examples of this literature for one country.

31. Regarding discrepancies between biblical authority and philosophical speculation, especially in such matters as the creation of the world vs. its eternity and the immortality of the soul, see below, n 47.

32. Aristotle, *Physics* 8.4; 235b14.

33. For a recent survey of these developments, see Edward Grant, *Physical Science in the Middle Ages* (New York, 1971).

34. Leon Battista Alberti, *I libri della famiglia*, ed. Ruggiero Romano and Alberto Tenenti (Turin, 1969), 86. Eugenio Garin, *Scienza e vita civile nel rinascimento italiano* (Bari, 1965), xiv–xvi, considers Alberti an important example of the sharing of scientific interests by humanists.

35. Joannes Ludovicus Vives, *De tradendis disciplinis* 4.6, in *Opera omnia*, ed. Gregorius Majansius (Valencia, 1785), 6:373 ff. [reprinted London, n.d.]. English translation by Foster Watson, *Vives: On Education* (Cambridge, 1913), 208 ff.

36. *De tradendis disciplinis* 4.1; Watson's trans. 166 ff., with slight change. Cf. *Opera omnia*, 6:347. Vives is hostile to all immersion in fruitless theoretical studies of nature; we should direct all our studies "ad vitae necessitates, ad usum aliquem corporis aut animi, ad cultum et incrementa pietatis." [Toward the necessities of life, toward some use of the body or soul, toward the cultivation and increase of piety] (ibid.).

37. Erasmus to Andreas and Christoph of Koenneritz (Agricola, ed. 1657, 679): "Visus sum mihi valles illas et colles, et foedinas et machinas non legere, sed spectare. Nec multum abfuit, quin ex tot venis argentariis et aurariis conceperim aliquam ejusmodi rerum cupiditatem. Utinam animis eo studio feramur in coelum quo scrutamur terram, non quod improbem hanc industriam, nobis enim terra gignit quicquid gignit, sed quod hae venae quantum vis faecundae, beatum hominem adeo facere non possunt, ut non paucos operae et impendii paenituerit, sola divinarum literarum vena vere locupletet hominem." [It seemed to me that I saw those vales and hills, those mines and machines, rather than read about them. So many veins of silver and gold all but engendered in me a desire for things of that sort. If only our mind carried us heavenward with the same zeal with which we search the earth. Not that I disapprove of this activity—for whatever the earth produces, it produces for us—but because however rich these veins may be, they cannot make man entirely happy: some few have regretted the labor and the cost. Only the vein of sacred books can truly enrich man.] The German translation of *Bermannus* carries the letter on 59–60. For the history of the letter, see Wilsdorf, *Georg Agricola*, 187–88.

38. Quoted from Lytton Strachey, *Eminent Victorians* (Garden City, N.Y.: Garden City Publishing Co., n.d.), 220.

39. The idea that only God will make the work succeed is often expressed by

Paracelsus, e.g., *Seven Defensiones*, trans. C. L. Temkin (in *Four Treatises of Theophrastus von Hohenheim Called Paracelsus*, ed. Henry E. Sigerist [Baltimore, 1941], 30): "Wherefore it follows from this that to those who walk in the way of God, perfect works and fruits grow of their talents which God gave them." This conforms to the Christian warning against the pride and haughtiness of those who ascribe their successes to their own prudence and virtue; cf. St. Ambrose, *Cain and Abel* 1.7.25, in *The Fathers of the Church: A New Translation* (New York, 1961), 42;384.

40. Laurentius Phries, *Spiegel der Artzney . . . gebessert . . . durch Othonem Brunfels* (Strasbourg, 1529), fol. viii^r.

41. Hieronymus Fracastorius, *De causis criticorum dierum libellus*, in *Opera omnia*, 2nd ed. (Venice, 1574), fols. 48v, 49v, 50v.

42. Giorgio Vasari, *Le vite de' più eccellenti pittori scultori e architettori* (Nelle redazione del 1550 e 1568), ed. Rosanna Bettarini, commentary by Paola Barochi (Florence), 1:31: "Introduzzione di Messer Giorgio Vesari pittore Aretino alle tre arti del disegno cioè architettura, pittura e scoltura."

43. See Carlo M. Cipolla, *Clocks and Culture, 1300–1700* (New York, 1967), 50 ff.

44. *Homocentricorum, sive de stellis, liber unus*, chap. 1, in Fracastorius, *Opera omnia*, fol. 2v. For Fracastoro's homocentric theory see J. L. E. Dreyer, *A History of Astronomy from Thales to Kepler*, 2nd ed. (New York, 1953), 296 ff.

45. Seneca, *Naturales quaestiones*, preface to bk. 1. Ibid., 7.2.3 (2:230 of the Loeb Classical Library), Seneca queries "whether the world goes round while the earth stands still, or whether the earth turns around while the world stands still." Although this may refer to the daily rotations only, Seneca yet concludes: "It is a matter worth contemplating, so that we may know where in the order of things we are: whether we are allotted the laziest seat or the fastest, whether god moves everything around us or us [around everything]."

46. Grant, *Physical Science*, 25 ff., and Paul Oskar Kristeller, *La tradizione aristotelica nel Rinascimento* (Padua, 1962), 21, seem to disagree on the medieval meaning of "double truth." For the age of Copernicus, however, it seems conceded that the theory meant "that something can be more probable according to reason and Aristotle, while the contrary must be accepted as true on the basis of faith" (Kristeller, 21). Granting to the theory its role in the emancipation of philosophy and science from theology, Kristeller adds (21 ff.), "I don't believe that the theory of the double truth as such was a conscious expression of free thought . . . but it certainly paved the way for the free thinkers of a later epoch, especially for those of the 18th century, who abandoned theology and faith and took advantage of a tradition which had established purely rational research as an independent enterprise."

47. This seems now generally agreed upon.

48. Hall, in *The Nature of Scientific Discovery,* ed. Owen Gingerich (Washington, D.C.: Smithsonian Institution Press, 1975).

49. Copernicus, *De revolutionibus,* bk. 1, chap. 6, fol. 6v and chap. 8, fol. 7r.

50. Ibid. 1.10, fol. 9v. This passage has often been quoted (e.g., by Wightman, *Science in a Renaissance Society* 121, and Kearney, *Science and Change,* 99 ff.) usually to establish the link between Copernicus and Hermetic philosophy.

51. Edward Rosen, trans., *Three Copernican Treatises: The Commentariolus of Copernicus, The Letter against Werner, The Narratio prima of Rheticus* (New York, 1939), 121–27.

52. For details see Rosen's introduction, ibid., 34 ff.

53. Rosen, *Three Copernican Treatises,* 122.

54. Paracelsus, *Seven Defensiones,* 20.

55. The Revelation of St. John 21:1–2: "And I saw a new heaven and a new earth; for the first heaven and the first earth were passed away; and there was no more sea. And I John saw the holy city new Jerusalem, coming down from God out of heaven." The new world and the Kingdom of God together follow the destruction of the old world.

56. On beauty and goodness in the science of the time, see Boas, "Philosophies of science," 250.

57. Cf. the work by Norbert Elias, *Über den Prozess der Zivilisation.*

58. Quoted from Martin Kemp, "Dissection and divinity in Leonardo's late anatomies," *J Warburg Courtauld Inst,* 35 (1972): 202n. 11.

59. See Andreas Vesalius, *De humani corporis fabrica libri VII* (Basel, 1543), 161 ff.

60. *Ricordi,* no. 147. Quoted from *Francesco Guicciardini: Selected Writings,* ed. Cecil Grayson, trans. Margaret Grayson (London, 1965), 38. Guicciardini does not doubt God's justice, but he thinks that "His counsels are so profound that they are rightly called *abyssus multa*" (ibid., no. 92, p. 26).

61. For this story see Alexandre Koyré, *From the Closed World to the Infinite Universe* (Baltimore, 1957), and Thomas S. Kuhn, *The Copernican Resolution: Planetary Astronomy in the Development of Western Thought,* reprint (New York, 1959), chaps. 6 and 7.

62. Unfortunately, I am unable to trace the source to which I owe this useful expression for one of the main obstacles to historical simplifications and generalizations.

CHAPTER 6

Gall and the Phrenological Movement

In the minds of most people, *phrenology* is associated with the doubtful art of reading a person's character from the configuration of his skull. A smaller number will also know that Gall, the founder of this art, was a renowned neurologist and the godfather of the principle of cortical localization of mental faculties. Only a few, however, will remember that phrenology in the first half of the nineteenth century represented a widespread movement transcending by far the circle of scientists and psychologists and affecting philosophy, religion, education, and literature. To indicate the breadth of the movement, it may suffice to state that men like Broussais, Bouillaud, and Abernethy among physicians,[1] Auguste Comte and Herbert Spencer among philosophers,[2] Balzac, Baudelaire, George Eliot, and Poe among literary men,[3] Horace Mann among educators, and Cobden among political writers[4] showed definite leanings toward phrenology. The complete history of this movement has still to be written.[5] The aim of the present chapter is to indicate its complex character and to show that this complexity was rooted in the teachings of Gall himself, which reveal two distinct aspects. A brief biographical introduction will sketch the origin of Gall's endeavors, the next part will deal with Gall's naturalism, and the last part will discuss the metaphysical element of his thought. Relatively little attention will be paid to the strictly phrenological (i.e., cranioscopic and organological) aspect of Gall's work, since this part is best known and has received competent consideration.[6]

→> • <+-

Originally published in *Bulletin of the History of Medicine* 21, no. 3 (May–June 1947).
© The Johns Hopkins University Press.

Franz Joseph Gall was born on March 9, 1758, in Tiefenbrunn, in the German state of Baden. He studied first in Strasbourg and then from 17○1 on in Vienna, where he obtained his M.D. degree in 1785 and set up in practice. In 1791 he published *Philosophischmedicinische Untersuchungen über Natur und Kunst im kranken und gesunden Zustande des Menschen*. This work is important because it expresses views on the relationship of body and soul and of man and his surroundings that show the early formation as well as unity of Gall's basic beliefs. It will, therefore, be discussed in more detail in the appendix to this chapter. The genesis of Gall's phrenological ideas must, however, be looked for in a different direction.

During Gall's formative years, physiognomy (i.e., "the art of judging character and disposition from the features of the face or the form and lineaments of the body generally"—*Oxford Dictionary*) was in great vogue, largely due to the enthusiastic efforts of Lavater. The origin of Gall's work must be seen in the framework of this movement.[7] As a boy, Gall had been impressed by a correspondence between the talents of his fellow pupils and their features. Those for instance, who excelled by their memory showed prominent eyes.[8] During the 1790s, while practicing medicine in Vienna, Gall reverted to his earlier observations in a more serious manner. His attention became centered upon the formation of the skull, and he began to gather experiences that led him to renounce the psychological classification of his time and to develop his own method of observation.

> For many years he endeavored to discover external signs in the head, corresponding to the general powers spoken of by metaphysicians, such as perception, conception, memory, imagination, and judgment; but not being able to advance, and finding contradictions and exceptions without end, he compared great talents for music, mechanical arts, drawing, painting, dramatic acting, poetry, philology, mathematics, and metaphysics, and since he succeeded with respect to the intellectual functions, he also looked to the head for signs of the different characters.[9]

But, not satisfied with mere craniology and cranioscopy, Gall took a definite turn away from physiognomy in favor of cerebral physiology. The physiognomist reads man's features as *expressing* his character. He is, therefore, free to believe in a soul that pictures itself in the shape of the body, similarly to God's creation of man in his own image.[10] Gall, however, did not read the configuration of the skull as an *expression* of man's instincts,

inclinations, and talents. "The object of my researches is the brain. The cranium is only a faithful cast of the external surface of the brain, and is, consequently, but a minor part of the principal object."[11] In other words, Gall considered the human character and intellect as the function of an organ, or rather as the combined functions of many organs localized on the surface of the brain. Thus, it is understandable why physiognomy, suitably associated with art, could be cultivated during the period of romanticism and "Naturphilosophie" in Germany, whereas Gall's doctrine was considered materialistic and was violently opposed by the idealistic natural philosophers in his own country.[12]

The accusation of endangering morality and religion was leveled against Gall in 1801–2 when the government forbade the private lectures he was wont to deliver in Vienna before a numerous and highly fashionable audience.[13] These lectures had made Gall's teachings known long before he published any comprehensive exposition of his doctrine.[14] It had been at one of these lectures in 1800 that Johann Caspar Spurzheim (1776–1832) had first become acquainted with Gall's thought. Spurzheim was then studying medicine in Vienna, where he had gone in 1799 after the French invasion of Treves had put an end to his theological studies.[15] According to Spurzheim, it was only after 1800 that Gall started upon his anatomical researches on the structure of the brain, first with the assistance of a young student, Niclas,[16] and from 1804 on with the very active help of Spurzheim,[17] who for about eight years became his close coworker. These anatomical labors were the subject of a memoir that Gall and Spurzheim in 1808 submitted to the Institut de France.[18]

Gall's reputation as an anatomist was very great in his day. He did not prove the existence of morphologically different areas that would correspond to the cerebral organs his doctrine assumed. But he presented a picture of the brain that did not contradict his physiological assumptions, and he established some facts that have remained of permanent value.[19] In contrast to most of his contemporaries, Gall did not dissect the brain by slicing it from top to bottom and thereby destroying the connection of its various parts.[20] Instead, he tried to show the course of nerve fibers and their relationship to the various gray masses from which they derived and in which they ended (as he believed). He claimed to have demonstrated "that the convolutions of the brain are nothing but the peripheral expansion of the fascicles of which it is composed: consequently, the convolu-

tions of the brain must be recognized as the parts where the instincts, sentiments, penchants, talents and, in general, the moral and intellectual forces are exercised."[21] On the other hand, the anatomy of the brain did not show any convergence of all fibers to one point, which might suggest the seat of the soul.[22]

In 1805 Gall left Vienna and, accompanied by Spurzheim, set out on a lecture tour through Germany and adjacent countries.[23] His success was sensational, though the comments were not all favorable and included attacks against his doctrine as well as his personal character. In 1807 he reached Paris as a man of international fame. In this city he remained to the end of his life, and here he published his main and most comprehensive work, the *Anatomie et physiologie du système nerveux en général, et du cerveau en particulier, avec des observations sur la possibilité de reconnoitre plusieurs dispositions intellectuelles et morales de l'homme et des animaux, par la configuration de leurs têtes.* The work comprised four large folio volumes and was accompanied by an atlas. The first two volumes (1810 and 1812) bear the names of both Gall and Spurzheim, whereas the remaining two volumes (1818 and 1819) have Gall as the sole author. While the second volume was being prepared, Spurzheim had departed for England to spread the doctrine in that country.[24]

This departure was the beginning of a rift between the two men. Spurzheim began to publish under his own name, deviating in many points from Gall's concepts. He gave a new name to the whole system, calling it *phrenology*, a name that he derived from the Greek word *phrēn* for mind and defined as "the doctrine of the special phenomena of the mind, and of the relations between the mental dispositions and the body, particularly the brain."[25] *Phrenology* has become the commonly accepted term for Gall's and Spurzheim's work in spite of the former's reluctance. When in 1822 Gall began to bring out a revised and handier edition of his main work, he did not adopt the name *phrenology* but gave it the title *Sur les fonctions du cerveau et sur celles de chacune de ses parties.* This revised edition was Gall's last contribution to his science. He died in Paris in 1828, when the reputation of phrenology was beginning to pale among experimental physiologists.

During Gall's lifetime, Flourens had begun his well-known experimental studies that assigned regulation of locomotor movements to the cerebellum, the respiratory center to the medulla oblongata, the sense of sight to the corpora quadrigemina, and intelligence as a whole to the hemispheres as a whole. On the other hand, Gall had found a supporter in

Bouillaud. As early as 1825, he had postulated the existence of a motor speech center in the anterior part of the brain, a postulate that was to be confirmed by Broca in 1861. Outside the circle of physiologists, however, the phrenological movement was still very strong and some of its most important manifestations were still to come. The movement did not rest on the outcome of scientific experiments alone, but to a higher degree on the vitality of certain philosophical principles. The best proof for this thesis is given by Flourens himself. As commonly assumed it was Flourens's book against phrenology, published in 1843, that proved fatal to phrenology. In this book Flourens's experimental results figured as a weighty argument, but they were an argument only to his main purpose, namely, to refute the philosophy of Gall and his followers.[26] The book was dedicated to Descartes, in whose name Flourens defended the freedom of the human will and the unity of the soul. It is with some of the philosophical principles as propounded by Gall and utilized by his adherents that the following pages are concerned.

Naturalism

Gall spent his life in two countries that were then separated not only by political barriers and linguistic differences but also by the philosophical outlook of their scientists. While Gall was laying the foundations of his system in Vienna, Kant's influence was in the ascendant and, when Gall finally left Germany, the idealistic schools of Fichte, Schelling, and Hegel were flourishing and "Naturphilosophie" was in full sway. From this movement Gall dissociated himself emphatically. As early as 1798 he asked to be excused "for not making use of the language of Kant" and ironically remarked upon his inaptitude "to arrive at knowledge a priori."[27] In his later works Gall repeatedly pronounced himself against speculative philosophy, and in no uncertain terms.[28] The belief that the world could be constructed out of the ego seemed to him as the negation of true science,[29] and Gall's concept of science can hardly be described as other than positivistic.[30] He stressed his positivistic attitude with regard to the basic tenet of his whole system,[31] namely the inborn nature of man's psychic and mental faculties and their dependence upon his somatic organization (i.e., his brain). Questions pertaining to this subject had hitherto not led to a conclusive answer for lack of sufficient facts. His own method would

be different. "Today we intend to put the phenomena in an orderly fashion and to expose their causes by demonstrating them by positive facts (*faits positifs*). We shall not present mere opinions but established truths which will at all times be capable of undergoing the proof of experience and which, consequently, must be of permanent usefulness for society."[32]

Gall was very anxious to keep his definitions within the limits of possible experience. To say that the *exercise* of the faculties of soul and mind depended on material conditions did not mean that these faculties were a *product* of man's organization. "We consider the faculties of the soul only insofar as, through the medium of material organs, they become phenomena for us. And without venturing beyond the material conditions, we deny and affirm only what can be judged by experience. We carry our researches neither on the inanimate body nor on the soul alone, but on the living man, the result of the union of soul and body." Whatever the decision of theologians and metaphysicians regarding the ultimate connection of body and soul may be, "our principle, to wit that the qualities of the soul and of the spirit are innate and that their manifestation depends on material organs, cannot thereby suffer the slightest alteration."[33]

This principle constituted a necessary premise for Gall's physiology of the brain, and he had to uphold it against two antagonistic doctrines. Phrenology was to be the natural science of man as a moral and intellectual being. But any such science was impossible if man's psychic qualities had no direct relationship to his organs, or if these qualities were but the accidental product of eternal things and our external senses.[34] In other words, Gall's defense had to be directed against spiritualists as well as extreme sensualists.

Of course, few if any spiritualists of the time went as far as to deny entirely any relationship between soul and body, since this would have led to absurd consequences. Moreover, Gall, as remarked, did not claim that the brain or its divisions produced these qualities. The cerebral organs were but the instruments through which the basic qualities manifested themselves. Indeed, Gall was far from being a materialist in the strict sense of the term. But he elaborated the old concept of the instrumentality of the brain in a manner that seemed destructive of man's spiritual entity and of his freedom of action.

There was, in the first place, his insistence upon man's relationship to animals. With many contemporary biologists Gall believed in the chain of

beings,[35] where the higher forms included the characteristics of the lower ones. However, he gave this principle an orientation that was far from common. Man and animal are subject to the same laws of organization, for the latter, too, have their psychic qualities represented in cerebral organs. The sexual instinct, love of progeny, attachment, the instincts of self-defense and of killing (including the penchant to murder), cunning, the desire to possess things (including the penchant to theft), pride, vanity, circumspection, memory for facts, the sense of locality, memory for persons, memory for words, the sense of language, the sense for the relationship of color, of musical tones, and of numbers, and the mechanical sense are shared by humans with many animals,[36] and the differences are of degree, not of essence. What then gives to man his decisive superiority over animals? Gall scoffs at the ready-made answers that animals are mere machines, devoid of all feeling and of moral and intellectual principles, man alone possessing an immaterial substance endowed with will and reason, or that the same qualities that in animals are ascribed to material forces or a vital principle in man are attributed to the soul acting independently from the body.[37] Gall proposes to solve the problem on the basis of comparative anatomy and physiology of the brain. The functions of the lower anterior part of the brain are much more highly developed in humans than in animals. But the upper anterior part of man's brain does not exist in animals at all. Consequently, an analysis of the organs of this region must yield the specifically human qualities.[38] As a result, Gall supplements his organological list with comparative sagacity, the spirit of metaphysics, wit, poetic talent, goodness, imitative faculty, religious sentiment, and firmness.[39]

It seems an imminent feature of philosophical systems, which see in man more than a natural being, to establish an inseparable gulf between animal and man. In Gall's scheme of moral and intellectual qualities, such a schism does not exist.[40] Humans possess several qualities that animals lack—but none of these qualities is of a radically different nature from those shared by both. Humans do not even show a special organ representing consciousness, intellect, reason, or will, nor is there any such arrangement as instinct versus intellect. Gall does not recognize "intellect" as a special entity at all. Potentially, there are as many "intellects" as there are qualities. Each of the latter can run the whole range from a mere instinct to a highly conscious and intelligent faculty.[41] Besides, several or

even all organs can function at the same time in a cooperative manner which, even in animals, can lead to very complicated mental acts and which, in man, explains the phenomena of reason and will. "Reason," meaning the ability to recognize general laws and principles, is "the result of a happy development of all anterior-superior parts of the brain."[42] Since these parts exist in humans only, they alone can have reason and are thereby distinguished from animals. But few men are endowed with reason in the philosophical sense given to this word, since in a few only are *all* the specifically human organs well developed. Gall remains in the sphere of biology: the chance of becoming a philosopher depends upon the development of the brain.

A somewhat similar thought underlies Gall's theory of will and moral liberty. From the very outset he denies the possibility of unlimited or absolute moral liberty, which would allow man to act without motive. "Moral liberty is nothing but the faculty of being determined and determining oneself by motives."[43] Man has no power over the existence of his desires and inclinations, which depend on his organization and the circumstances stimulating it; he is obviously determined by internal and external motives. But this accounts for only the first half of the definition. Because man is endowed with a multiplicity of cerebral organs of a high order, he can choose among the motives and thereby determine himself. An animal that had one cerebral organ only would obviously be limited to one single kind of sensation and one idea and would, therefore, have no range of choice. In higher animals that possessed several organs, matters were already different. Supposing that a hungry dog were shown food, but that a hare was made to run in front of him just when the dog was on the point of eating. Still hungry, he would set out after the hare. However, if he had repeatedly been prevented by force from giving chase to the hare, he would remember the beating that awaited him and would refrain from the pursuit. "If the dog were susceptible to hunger only, or if his sole penchant and faculty were for hunting, this manner of acting would be impossible to him. Thus it is the plurality of organs which makes him susceptible of different ideas and sensations." Yet this is no moral liberty, "this faculty must be considered as a simple spontaneity or faculty of being determined by the strongest and most numerous stimuli."[44] Or, to use modern terminology, the animal may be *conditioned*, but it will not have any will of its own. Only in man does "will" really exist, since his supe-

rior qualities allow him to distinguish truth from error, justice from injustice, and to possess a foreknowledge of future events, sympathy, and
compassion. If, for instance, man is provoked to yield to the desire of
vengeance, he knows "that a base action will dishonor him and that he
will be considered the slave of his passions rather than the master of himself. If he throws himself into the arms of pleasure, the frightening image
of his destroyed health and of his ruined domestic felicity presents itself
before his eyes. The social conveniences, the shame of abusing confidence,
the grievous consequences of his conduct for the beloved object, etc., all
these motives act on his mind and by their force or by their number finally
get the upper hand."[45]

As the last words show, even in the realm of human moral action the
decision will yet fall on the side of the strongest motive. It may, therefore,
well be asked whether Gall's definition of moral liberty has any justifiable claim to this designation. The spiritualists among his enemies certainly denied it: "Liberty is precisely the power to determine against all
motive," Flourens objected.[46] It may also be asked whether Gall's concept
of moral liberty would allow phrenology to be "grafted upon Kant's system" quite as easily as has been contended.[47] It is not the categorical imperative that Gall opposes to the motives, not man's intelligible nature
against his empirical character. By necessity man decides "according to the
motive which acts most powerfully upon him or which offers him the
greatest good."[48] Because of this necessity, man's moral and social life still
remains in the confines "of the laws of nature and of our organization."[49]
This in turn enables Gall to give a rather utilitarian interpretation of social institutions and to say that "education, morals, laws and religion are
indispensable auxiliaries for assuring the happiness of man since these institutions supply him with the most numerous, most powerful and most
noble motives that can induce him to act well."[50]

At this point the question arises as to how Gall could have any trust
in education or any institution that aimed at improving man. It would
seem that, if man's faculties for good and bad were innate and, moreover,
dependent on the constitution of the brain, his character must be fixed
from his very birth.

The problem of education was indeed a crucial one, which Gall had
met very early in his development. When his ideas and observations about
the craniological differences among humans had begun to crystallize, he

had distrusted them. "A great number of philosophers and physiologists contend, I told myself, that all men are born with equal faculties and that the differences which one notices among them were due either to education or to accidental circumstances. If this is so, there cannot be any external signs of the dominant faculties and consequently the project of learning in this way how to know the functions of the diverse parts of the brain is a veritable folly."[51] The strong belief of the Enlightenment in the natural equality of man and the possibilities of education is too well known to need any elaboration. In his later work Gall chiefly pointed to Helvetius (1715–71) as the main protagonist of the doctrine that education was all powerful in shaping man's character.[52] Gall had overcome these objections by observing that children brought up in the same environment still showed marked differences and that the teachers themselves obviously heeded the differences among their pupils and merely admonished them to develop the gifts God had given to them.[53] Thus Gall defended the inborn nature of man against the radically sociological approach. Man's nature did not change. It differed somewhat according to nationalities and individualities, but otherwise it was the same at all times.[54] But he couldn't blind himself to the fact that human culture had manifested itself differently in the course of history. Hence he was led to believe that the inborn nature of man and his surroundings combined in shaping his *manifest* behavior. The external influences stimulated the action of the inborn faculties and supplied the necessary motives.[55] Moreover, this form of stimulation could be used to influence man's organization. The cerebral organs, like other organs of the body, were strengthened by use and weakened by disuse. Hence, moral, religious, and educational efforts had to stimulate the human organs and carefully avoid stimulation of the animal ones.[56] In other words, education could be effective if it was based on a scientific study of man's organization (i.e., phrenology).

Gall was especially interested in the treatment of criminals, and during his lecture tour in Germany he used every opportunity to visit the jails, to examine the heads of the inmates, and to compare their records with his findings. He has been rightly called a forerunner of Lombroso's idea of the born criminal.[57] With equal justice he might be called a forerunner of the idea that crime is a medical problem and the criminal a victim of his disposition and an object for well-considered mental treatment. He did not go so far as to deny moral responsibility in all cases. Moreover, he

asserted that punishment was a very potent means in motivating the potential malefactor to suppress his natural inclination.[58] But he emphasized that moral responsibility varied from case to case.[59] And his whole pragmatic concept of moral liberty, as well as his analysis of many examples cited, suggested that crime was the result of the preponderant development of "animal" organs, of downright mental alienation, and of a lack on the part of society of providing the criminal with sufficiently strong motives of a higher order.[60]

In 1807, the year in which Gall settled in Paris, French medical scientists were largely under the influence of a philosophical movement to which sensualism gave the dominant note.[61] Condillac had deeply influenced the physicians Bichat, Cabanis, and Pinel, and Destutt de Tracy was presenting a new synthesis of their thought under the name of "idéologie." According to Destutt de Tracy, intellectual phenomena were but modifications of feeling, of which he distinguished sensibility, memory, judgment, and will. Cabanis had given the physiological basis for this theory; according to him thought was produced by the brain "just as the stomach and intestines are destined to effect digestion, the liver to filter the bile, and the parotid, maxillary and sublingual glands to prepare the salivary juices." But Cabanis had made an important step beyond Condillac. Whereas the latter's attention had centered upon the sensations provided by the external senses, Cabanis had recognized the importance of internal sensations originating in the inner organs. These internal sensations accounted for instincts and passions. Similarly, Bichat had assigned the passions to the organic life as distinct from the animal life, which latter was dominated by the brain. Neither Bichat nor Cabanis, however, expressed himself with unmistakable clarity regarding the relationship of all psychic phenomena (i.e., including instincts and passions) to the central nervous system. Cabanis still adhered to the doctrine of "temperaments" as shaping man's personality.[62] Moreover, he and Bichat recognized the existence of imperceptible sensibility as independent from conscious sensations.[63]

This sensualistic movement led by the ideologues was already being decried as "materialistic" by theological traditionalists when Gall came to Paris. But Cabanis was still alive, and the truly philosophical opposition had not yet formed. Surely, then, here was a climate congenial to Gall's

doctrine. The animal in man, the brain as the organ of the soul, moral institutions as "conditioning" humans for correct behavior were concepts that would hardly frighten the leaders of French scientific thought. They might hesitate to accept the details of Gall's organology and might reject his cranioscopy, but they were not likely to attack him on spiritualistic grounds.[64] Indeed, Gall and the sensualists seem to have so many points in common that Gall's name has repeatedly been associated with them.[65]

This view is not without justification. Sensibility, according to Gall, is the most general property of the nervous system.[66] But in the higher animals and in humans the whole nervous system is under the dominance of the brain, and sensibility is inseparable from sensation (i.e., the perception of a stimulus). Sensibility without consciousness is a contradiction in terms; the multitude of reactions that go on in our body without our being aware of them have to be explained by irritability.[67] By assigning a preponderant role to the brain, Gall exerted considerable influence upon contemporary medicine. The brain now was considered the unique organ of all higher functions, including instinctive urges and passions, a view particularly important for the psychiatrist and accepted even by many of Gall's adversaries.[68]

But apart from these physiological corrections, Gall criticized sensualism in a manner calculated to give it an entirely new turn. In the first place, he denied Condillac's contention that we receive our ideas and knowledge through the external senses alone. "In that case man and animals are the perpetual toy of external, fortuitous and changeable objects; the measure of the faculties has no other base than the perfection of the senses; and education whose goal should be to make of individuals and of nations what one wishes has no other secret than to calculate duly the action of the outside on the senses."[69] In reality, what we think and desire is the combined result of the organization of our brain and the impressions from outside. Mere sense perceptions cannot even account for our intellectual acts. In the second place, Destutt de Tracy, according to Gall, had not grasped the true elementary qualities of our intellect. To say that memory, judgment, and will were modifications of feeling was to deal in abstractions. These modifications were but attributes common to all primary qualities and functions. The task for the physiologist was to find the latter.[70] Gall did not dislodge sensation from its place, but he declared it a kind of common factor that, once admitted, could be neglected. Instead,

he concentrated on the discovery and explanation of the different forms of sensation (i.e., the twenty-seven primary qualities and faculties and their cerebral organs). In doing so, he had shifted from an introspective and psychological analysis to a descriptive and physiological investigation.[71]

Destutt de Tracy had declared ideology a part of zoology.[72] But his analysis of sensation was based on an attempted insight into the working of his mind and, therefore, was quite different from the method a zoologist would use in studying the behavior of animals. This latter method was, however, the one followed by Gall, who observed animals and his fellow men, compared their various instincts, talents, and deficiencies, and tried to locate them in the brain. He, therefore, came much nearer to the physiological ideal of the ideologues than they themselves had done.

This was soon to be made manifest by historical development. During Gall's lifetime, the ideological school steadily declined. Toward the end of the empire, and ever increasingly in the subsequent years, the philosophical opposition against the *idéologie* gathered momentum. First the Scottish philosophers of the "common sense" school were introduced,[73] then the Germans Kant, Fichte, and Schelling were studied with increased interest, and Plato and Descartes, with his dichotomy of body and soul, again attracted attention. A new school of "eclectics" or "psychologists" came to the fore, and by 1830, under the leadership of Victor Cousin and Théodore Jouffroy, this had become "the official philosophy."[74] Introspective psychology as the science of the mind and as a steppingstone to metaphysics was one of the main features of this new school. Whatever the vicissitudes of its doctrines may have been otherwise, "Sensualism" was charged with materialism, atheism, and "revolutionary violence and servility."[75]

The ideologues proved rather helpless. The inherent contradiction of their system (i.e., their failure to give a satisfactory physiological interpretation of the vast world of ideas) was bound to lead to a split. The choice lay between psychology and physiology. Maine de Biran (1766–1824), once the bright hope of Cabanis and Destutt de Tracy,[76] had taken the road in the first direction. Physiologists like Magendie, who had been reared in the teachings of the ideologues, put their trust in empiricism, growing increasingly disdainful of all theories. Thus, the field seemed to be clear for eclecticism when, in 1828, Broussais's book *De l'irritation et de la folie* appeared, which carried a blistering attack against the "Kanto-

Platonists," "psychologists," and "spiritualists." These men, Broussais said indignantly, were trying to do away with the philosophical achievements of Bichat, Cabanis, Destutt de Tracy, Pinel, and others who had followed the path of objective science. Instead, they tried to impose the foreign concept of man as an intellectual and moral entity that was to be studied by the science of the consciousness. This was a dangerous and impossible undertaking. The phenomena of consciousness are not sufficient to constitute a science. The words *reason, ego, consciousness* "merely express results of the action of the nervous matter of the brain, an action which is susceptible of change as long as the state of life lasts."[77] *Irritability*, defined as the property of living tissues "to move upon the contact with a foreign body," is at the bottom of all this.

> When man has the consciousness of the movements excited by the foreign bodies . . . one says that he has felt the impression of these bodies and one gives the name of sensibility to the faculty which he has of feeling them. Thus sensibility belongs to the *ego* and irritability belongs to all the fibres of man's body. A part affected by the foreign bodies may experience movements without the *ego* being conscious thereof. In this case there is irritability only. If, however, the *ego* experiences a modification which makes man say *I feel*, there is irritability and sensibility. Sensibility, therefore, is the consequence of irritability, whereas irritability is not the consequence of sensibility.[78]

The inner world of consciousness exists, but since it is dependent on external conditions it cannot be made an independent science. Broussais knows that he is unable to explain how thought is produced by an irritated nervous system. But this is no reason for substituting a hypothetical entity as the spiritualists do. The whole difference between physiologists and spiritualists can be summed up as follows: "We have proved by the facts that all the instinctive and intellectual phenomena are acts of the irritability of the nervous system, but we have refused to explain the how. We essentially distinguish the fact of the production of thought by the brain from the explanation of this fact, whereas the spiritualists deduce the impossibility of the fact itself from the impossibility of its explanation and place an entity in the brain in order to supply this explanation. We reply that this entity is a hypothesis."[79]

The book made a great stir and caught the eye of friend and foe. It is

said to have earned Broussais his seat in the "Institut,"[80] in the section of philosophy, a successor to the "section d'analyse" that the ideologues had once dominated.[81] The eclectics on their part also acknowledged the importance of this book, with Damiron devoting a large chapter to Broussais as one of the representatives of the sensualist school. Broussais tried to vindicate Cabanis and his circle, even including his old antagonist Pinel. He had attacked Pinel because the latter, with his "essential fevers" and his classificatory system of diseases, had represented ontology in medicine. Pinel had considered diseases as autonomous entities; Broussais conceived them as mere results of abnormal irritation of tissues. These differences had been medical. Now, before the threat of spiritualism, Broussais felt himself philosophically united even with Pinel.[82] But his uncompromising attack against the new metaphysical ontology did not solve the difficulty inherent in ideology. The criticism was negative; he had no positive doctrine to oppose to introspective psychology. Yet, as a matter of fact, such a doctrine existed in Gall's system; it was but necessary to accept it. Sometime after 1828 Broussais actually made this step and became an enthusiastic phrenologist.

It is not without interest to observe Broussais's changing attitude toward Gall. In 1821 Broussais was still critical of Gall, making light of the idea that passions and instincts, just as intellectual faculties, had their center in the brain.[83] In *De l'irritation et de la folie* of 1828, Broussais's tone has changed. Gall has acquired "eternal rights to the gratitude of men" by his great work on the functions of the brain.[84] "The foundations of his system are very solid; we regard him as one of those who have best understood the functions of the nervous system and we are indignant at the levity and ingratitude with which writers have treated him who have hardly left the ranks of his hearers and who owe him anything reasonable they have said about the functions of the brain."[85] Nor does Broussais find any fault with the fundament of Gall's doctrine, "which consists in referring all intellectual and instinctive phenomena to the action of the encephalic apparatus."[86] Nevertheless he cannot yet follow him in his organologic system.[87] But by 1836 the conversion is complete; Broussais delivers a course of lectures on phrenology that is attended by such a large audience that a new hall has to be found. At the very outset Broussais insists that phrenology is not a psychological system but "the physiology of the brain." Gall has felt that observation of oneself is not sufficient and that it is neces-

sary to study others. Introspection reveals man in certain situations only, whereas observation of others reveals man in a much broader light. "What follows from this? that the study of man is an observation of natural history and that then psychology is no more than a branch of it, no more than the history of some of the circumstances in which man may find himself."[88]

Broussais's deviations from Gall can here be neglected. What matters is to show that his final acceptance of phrenology represented a logical historical development (i.e., logical from the point of view of his physiological premises).[89]

But as far as Broussais personally was concerned it seems that logic was helped by a human agent, the philosopher Auguste Comte. When Comte, in 1854, published the fourth and last volume of his *System of Positive Polity*, he added an appendix containing his early essays. The last of these essays, originally printed in the *Journal de Paris* of August 1828 and now reprinted, was a review of Broussais's *De l'irritation et de la folie*. Commenting on this essay, Comte now said: "The insight gained through my personal experience was utilised in this review of the memorable work in which Broussais worthily combated the metaphysical influence. This concluding Essay will ever possess an historical interest since it roused the great biologist to the noble effort which produced, at the close of his admirable career, his just appreciation of the masterly conception of Gall, till then disregarded by him."[90]

Whether Comte really converted Broussais to phrenology or whether he merely flattered himself of having exerted this influence, so much at any rate is certain: to Comte's mind, phrenology was the culminating point in the positive study of man as a natural being. As early as 1825, with clear reference to phrenology, he had said, "All who are truly on a level with their age, are aware that physiologists, in our day, study moral phenomena exactly in the same spirit as other phenomena of animal life."[91] In his review of Broussais's book, Comte had been more explicit. Following the other sciences, physiology had taken the turn toward positivism in the second half of the eighteenth century. But to make the development complete, the intellectual and moral phenomena also had to be included.

> Thus the memoirs published at the beginning of this century by Cabanis on the connection between the physical and moral nature, are the first great and direct effort to bring within the domain of positive physiology this

study previously abandoned entirely to the theological and metaphysical methods. The impulse imparted to the human mind by these memorable investigations has not fallen off. The labours of M. Gall and his school have singularly strengthened it, and, especially, have impressed on this new and final portion of physiology a high degree of precision by supplying a definite basis of discussion and investigation.

Needless to say that Comte praises Broussais for his fearless attack upon the psychologists and for having shown the futility of introspective analysis. But Broussais has not gone far enough, and Comte blames him for not having "proved that such internal observation is necessarily impossible." Comte makes this impossibility clear on the basis of a phrenological argument: the observing organ cannot observe itself directly. Another point on which Broussais is blamed concerns his failure to make clear "the immense difference which exists between the physiological doctrine of man, intellectual and moral, and the theories of the metaphysicians of the last century, who saw in our intelligence only the action of the external senses, disregarding every predisposition of the internal cerebral organs."[92] He particularly refers to Condillac and Helvetius, both of whom Gall had attacked. In his *Cours de philosophie positive*, Comte also includes Destutt de Tracy, who had declared ideology a part of zoology, only to treat it right away as an independent and primary science. Gall and Spurzheim, he says here, have refuted "the puerile reveries of Condillac and his successors on the *transformed sensation*." [93] Unlike Broussais, Comte had not graduated from the sensualist school. An erstwhile collaborator of Saint-Simon and a sociologist rather than a physiologist, Comte was able to view the *idéologie* much more objectively and to see that Gall's criticism of sensualism and ideology was justifiable on the basis of the latter's own premises.

Phrenology played a very important role in Comte's system,[94] as evidenced by the position he assigned to it in his *Cours de philosophie positive*. It forms the subject of the forty-fifth lesson, which he wrote in December 1837 and which is entitled "Considérations générales sur l'étude positive des fonctions intellectuelles et morales, ou cérébrales." Here Comte proves himself by no means a blind follower of Gall. On the contrary, he very clearly sees the many shortcomings of the doctrine and emphasizes his concern with its basic concepts rather than its details. The basic concept he regards as correct and as constituting one of the principal elements of

the philosophy of the nineteenth century. As to the details, particularly Gall's table of primary functions and the way he localized their organs, he contends that it was better to establish a hypothesis that could afterward be verified than to refrain from giving the system a form of completeness necessary for its propagation.[95] This was essentially the same stand as taken by Comte regarding Broussais's reform of pathology. Broussais had conceived of disease "as essentially consisting in the excess or deficiency of stimulation of the different tissues, either rising above or falling below the degree which constitutes the normal condition. This conception throws a great light on the nature of diseases, by exhibiting them as results of mere changes of intensity in the action of stimulants indispensable for maintaining health." If Broussais had "erred as to the real seat of a particular malady, it was better for pathology, and even for therapeutics, to propose a seat at variance with the true one than none. M. Broussais has thus definitely led thinkers to the true road of observation, where, even while combating his ideas, they can only serve the progress of science."[96]

Thomas Huxley once expressed astonishment at Comte's lack of comprehension of the main features of science, "his strange mistakes as to the merits of his scientific contemporaries; and his ludicrously erroneous notions about the part which some of the scientific doctrines current in his time were destined to play in the future." Huxley certainly included Gall and phrenology in this statement.[97] But the example of Broussais shows that Comte was not as shortsighted as Huxley would have him. In Broussais's work Comte anticipated something of Virchow's famous definition of disease as life under changed circumstances. Broussais's mistakes, he felt, since they concerned questions of fact, could be left to verification. And the same was true of his attitude toward Gall. Where else but in Gall's system could he have found support for a complete positive system of science and for his "behavioristic" concept of man so important for his political outlook?[98] The keenness of Comte's sight is attested by Flourens, who combated phrenology lest Gall's philosophy might become that of the nineteenth century.[99] For Comte, the Positivist, and Flourens, the "psychologist,"[100] interpreted Gall alike, though they evaluated him quite differently. And in 1877 Ferraz could sum up the situation by saying that Gall and Broussais had formed the link between *condillacisme* and *comtisme*, which latter, as Ferraz stated, "had assumed such great authority under the second [empire] and which at this moment pretends to universal domination."[101]

Indeed, it can be said that even today Gall is largely seen through the glasses of his positivistic interpreters. Emile Littré in France and George H. Lewes in England, who were both strongly influenced by Comte, though they refused to accept all the tenets of strict Positivism, defended Gall's reputation at a time that was fast forgetting its indebtedness to the man over the laughter at his fantastic claims. Broadly speaking, Littré and Lewes agreed that phrenology was dead and that Gall's scheme of specific functions and cerebral organs had not survived the test,[102] but that he had led the right way in assigning the study of mental functions to biology[103] and in posing the problem of cerebral localization.[104] "Indeed all that relates to the general propositions respecting a plurality of functions, and a plurality of organs, Gall must be admitted to have triumphantly established. It is only in the details that he is unsuccessful."[105] Thus they molded the picture of Gall as a great, though misguided and perhaps even slightly ridiculous, figure in the rise of a progressive science.

Religion and Politics

In the preceding section I presented the naturalistic element of Gall's system and showed how this element entered into the positivistic movement, especially of France. The presentation emphasized those features that are mostly remembered today. The picture drawn is true enough as far as it goes, but it is not complete; it does not portray the historical Gall. The positivist Gall has to be complemented by the metaphysician Gall, who had his roots in a tradition different from the revolutionary sensualism of the late eighteenth century.[106]

Gall has rightly been called a deist. "To be an atheist and, moreover, even to profess it, has always seemed absurd to me." On the other hand, he was rather indifferent to dogmas and creeds and not above occasional cynical remarks in religious matters.[107] Although admitting that the soul of man was one and the same and the brain but the material basis for the exercise of the mental faculties,[108] he shunned discussions on the immortality of the soul "because this does not concern the physiologist directly."[109] Obviously, Gall was not a pious person in the orthodox sense.

Now it may well be asked why Gall, who pretended to avoid religious questions not pertaining to physiology, did not follow the same policy with regard to man's creation and destiny. Far from avoiding the problem

of creation, he repeatedly dwelled upon it and even offered a proof for God's existence based on the organ of religious sentiment.[110] The answer to the question is that, at Gall's time, and especially outside France, physiology had not yet severed all ties with natural theology. Physiologists not only used teleological expressions (as they still do today), but also attributed the seeming purposefulness of organic structure to design on the part of God. Indeed, from Galen's *On the Use of Parts* down to the Bridgewater treatises,[111] many physiologists believed that their science helped to prove the existence of God and a wise Providence. Gall clearly belonged with these men and sometimes expressed himself in an enthusiastic vein reminiscent of Galen.[112]

But Gall is not primarily intent on proving design. Rather, he takes design for granted. He remembers his school days, when his teachers often spoke of "natural gifts, of God's gifts," and admonished him and his fellow students "in the spirit of the Gospel, telling us that each of us would have to render an account in proportion to the talents he had received."[113] The conviction of the divine and immutable order of the physical as well as moral world runs like a leitmotif through his philosophy and serves as a justification for his basic principle that man's qualities and faculties are inborn.[114]

According to Gall it is wrong to assume, as the ideologists had done, that external needs shape animals and humans. On the contrary, the needs come from inside as the result of the action of the cerebral organs. Therefore, Lamarck is mistaken in believing that needs and usage molded organs or brought new ones into existence.[115] What is true in the realm of biology holds equally true in the social world. It is a mere dream to imagine some natural state of man to which good and bad qualities have been added by his social life. Some animals live in isolation, others in community.

> Man has been destined to live in common. Nowhere and at no time has he lived isolated. As far as one can go back in history, he has been united in families, hordes, tribes, peoples and nations; consequently, his qualities also had to be calculated for society. The phenomena one observes in whole peoples are no more the effect of this union than are the phenomena which take place in each man in particular. Always and everywhere has the human species manifested the same penchants and the same talents; always and everywhere this has led to the same virtues and the same vices, the same

industries and the same institutions. . . . The only changes which one notices in the progress of human society consist in that the same penchants and the same faculties are exercised upon different objects and produce modified results. The morals, customs, laws and different religious ceremonies of the different peoples rest upon the same bases. . . . Sing your love upon the reed-pipe or the harp, adorn your chiefs with feathers or the purple and your women with flowers or diamonds, live in huts or in palaces—it is always the same faculties that make man act in the circle which his Creator has drawn for him.[116]

The significance of this stands out most clearly in its consequences, which put Gall in contrast to "progressive" thought of his time. His attitude is thoroughly unhistorical. French physiologists like Bichat and Magendie had carefully evaluated the relative shares of natural endowment and civilized life in man's higher faculties. Of course, Gall did not deny the influences of cultural life or of education. But these influences can only develop or suppress, they can neither create nor abolish what is inherent in man. "Thus all the allegedly factitious qualities and faculties are the original appendage of the human species and not the subsequent effects of invention and of discoveries. It is in the disposition of men without culture, in the disposition of savages and barbarians, that one must study the natural dispositions of the civilized nations." In other words, the study of the primitives reveals man as he really is; it is anthropology in the true sense of the word. And the primitive in man always remains alive: "Study the penchants and faculties of civilized men and you will recognize the penchants and faculties of the savages and barbarians."[117]

The slogan "Human nature never changes" even today is often favored by the biologist skeptical of social utopias where man will be peaceful, good, and wise. This skepticism has a radical protagonist in Gall, who denies an ever-progressing perfection of mankind. "For the whole known duration of the human race, the generality of men has always been subjected to ignorance, error, prejudice and superstition. Slavery, brutality, the loud and gross gratifications of the senses have always been its share."[118] Hunters, fishermen, peasants execute everything mechanically and are hostile to all innovations. Craftsmen behave very similarly, and in the "higher classes" things are not so very different either. The rulers could exercise an enormous influence upon the arts and sciences, and they know

that by encouraging talent and industry they gain fame and the blessing of future centuries. But they succumb to the immensity of their task, to the tyranny of etiquette, and the jealousy and ignorance of their favorites. The idle rich squander their lives in the pursuit of idle honors and sensual pleasures. "Hence the sad truth that the class of men who from their imaginary height, look with contempt upon the populace is nevertheless on the same level as the latter as far as their lights are concerned."[119] Philosophers fall into extravagant errors, not to speak of the interminable controversies of theologians, the eternally vacillating forms of government, and the infantile state of criminal legislation.

Have poetry and the fine arts progressed? A look at the monumental buildings of the ancients and at the artistic and poetical works of past ages will be convincing argument against an ever-progressing perfection. "Whatever man can attain immediately through the energy of his faculties and qualities, all that is in the province of genius, he has attained and will attain whenever the development of his organs has been or will be favored to a very high degree by nature." The positive arts and sciences seem to form an exception. Favored by particular circumstances they can advance far beyond their initial state. But even they do not contribute to the moral or intellectual perfection of man. New knowledge is acquired at the expense of the old. Even if the mass of society is enriched by countless new discoveries, the individuals do not thereby become more impressive. Everybody has to restrict himself to a particular field so as to keep abreast. "And no sooner has one started to soar to the height of one's domain than one is thrown into the abyss of nothingness." The same is true of nations; Athens and Rome have fallen back into barbarism, and barbarians, on the other hand, have shown all the political virtues and vices to be found in the most modern states. In short, the history of all nations is the same and "the moral perfectibility of the human species is confined within the limits of its organization."[120]

From a moral point of view, the human race is neither good nor bad but both. Man is endowed with benevolence, religious feelings, and other good qualities. But he will also steal and murder, and there will always be wars,[121] since all these acts are founded in his natural desires. And since God has created man, one must recognize the existence of moral as well as physical ills as entering into "the plan of eternal Providence" and must submit, for the one as for the other, "to the decrees of God."[122]

Gall's metaphysical concept implies a strict harmony between moral and physical laws. Both have their source in God, and there can be no discrepancy between the two. Gall was only consistent when he said: "All man does, all he knows, all he can do and all he can learn, he owes to the author of his organization. God is the source thereof; the cerebral organs are the intermediate instruments which He uses." Driving this view to its very extreme, he exclaimed enthusiastically, "God and the brain, nothing but God and the brain!"[123]

Gall was subtle enough to see that this doctrine might easily lead to a fatalistic philosophy. In his own way he was faced with the age-old dilemma between the concept of an ordered universe and man's urge to see meaning in his actions. The theological antinomy between God's will and the moral liberty of his creature, the dialectician's difficulty of reconciling the historical process with the desire to play an active historical role, the scientific contradiction between man as a mere particle of nature and man as the ruler over nature are but different aspects of the same dilemma. By assuming the existence of inborn evil tendencies in man, Gall tried to solve the dilemma. Without the inclination toward evil, there would be neither vice nor virtue, and man's actions would be subject to neither praise nor blame. Here Gall's arguments are full of quotations from the Gospel and the fathers of the church, and indeed what Gall defends is the doctrine of original sin.[124]

The whole peculiar combination of science and theology as sketched in this chapter was irreconcilable with the creed of French liberals and Comtists. Comte himself rejected Gall's historical speculations by the same argument he had used against Cabanis, who had "confounded the study of individual man with that of the human race collectively regarded." The latter is the domain of sociology which, though based on physiology, is yet a separate science "requiring special observations on the history of the development of human society, and special methods."[125] And Gall's theological views he simply ignored.[126] The liberals, as represented by Broussais, likewise tried to evade Gall's theological preoccupation by suggesting that the latter's proof of the existence of God might have been motivated by convenience rather than belief.[127]

The French liberals were indeed in a difficult position. Phrenology was not only the climax of the *idéologie*, it was also a leaven of anticlericalism. At a time when Xavier de Maistre interpreted disease as an expression of

sin,[128] the phrenological fashion of the day interpreted crime as a disease and deplored great criminals as victims of their organization.[129] Thus, in the words of Damiron, "Some have seen in this system an antimystical, antitheological and antisacerdotal idea and have then elevated and defended it like a flag, they have vowed to keep its memory bright; others on their side have seen in it an impious and immoral doctrine which they have treated with violence and covered with maledictions."[130] Even after the phrenological system and Gall's work had been pushed into the background, the idea of cerebral localization was upheld as a liberal dogma, and in the sixties the discussions about the speech center were strongly tinged by materialistic and republican sentiments.[131]

Gall deeply believed in the search for truth as the noblest task of the scientist and deplored any obstacles put in the path of scientific research. It can also be admitted that Gall was more interested in the truth of his doctrine than in any theological or political consequences it might have. "My doctrine can exist side by side with any religion and any constitution."[132] But Gall was certainly not an antireligious person nor a democrat in politics. Characteristically enough, one of his main patrons was Prince Metternich,[133] and his medical practice in Paris extended among the most aristocratic circles, notably the members of the diplomatic corps.[134] Moreover, Gall's political indifference and theological metaphysics were supplemented by opinions where belief in the inequality of man mingled with downright contempt for the majority of men. "The study of the physiology of the brain shows us the limits and the extent of the moral and intellectual domain of man. It shows us an immense disproportion between the mediocre faculties and the eminent faculties, and it leads us to the conclusion that wherever regulations, decisions and laws are the work of the majority of votes, mediocrity triumphs over genius. *Propter peccata terrae, multi principes ejus.*"[135]

Of course, the whole system of cranioscopy rested upon the principle that men were born unequal as to their moral and intellectual faculties. But Gall developed this principle in a very extreme direction. The great majority, he held, was formed everywhere by mediocre men who by themselves invented nothing and created nothing. Only those with eminently developed organs that transmitted their inner stamp to the external world were really creative. Against Helvetius, Gall denied that every well-organ-

ized person could equal the very great.[136] The "greatness of the human race" was altogether a futile dream,[137] and Aristotle was right that some men were born to dominate, others to obey.[138] Certainly this was not the creed of a democrat or liberal. Fundamentally this was the attitude of a man who did not wish to upset the providential order of things but looked upon his fellowmen with slightly cynical benevolence and the conviction that in his science he had found an instrument by which they could be administered rationally as well as effectively.

In the Anglo-Saxon countries not less than in France, phrenology in the early nineteenth century was part of the "free-thought" movement and regarded with suspicion by the churches.[139] But in contrast to France, free thought here was allied with a deism that took itself seriously rather than with atheism or religious indifference.[140] Thus, in the Anglo-Saxon world, Gall's metaphysical and religious speculations were no hindrance to the popularization of his system in liberal circles. This only needed a person who would give greater emphasis and a slightly different direction to this side of phrenology, and this person was Spurzheim, the erstwhile student of theology. Spurzheim criticized Gall for having confused primary faculties, on the one hand, and character and action, on the other. As an empiricist, Gall started out with what he could observe directly, and so he believed in the existence of poetical, religious, metaphysical faculties and others of this kind that were manifestations of primary faculties in certain persons but not the faculties themselves.[141] Besides, Spurzheim distinguished between affective faculties (comprising penchants and sentiments) and intellectual faculties (comprising external senses, perceptive and reflective faculties), which he arranged in the systematic order of a botanist or zoologist. Spurzheim's scheme, which increased the total number of faculties to thirty-five, became more popular than Gall's classification. But the most important feature of Spurzheim's revision—in the present connection, at least—was the omission of any "evil" faculty. In deliberate contrast to Gall, Spurzheim maintained that "all faculties in themselves were good and given for a salutary end" and that by abuse only could the functions resulting from them become evil.[142] Thus Spurzheim softened Gall's pessimism; if all men had been created potentially good,

then there was no need to despair of the progress of mankind. Whereas in Gall's system the benefit that phrenology promised to confer was limited by the providential existence of evil, Spurzheim again could hope for the *perfection* of mankind.[143]

To Spurzheim, science, religion, and morality became one. Phrenology was the science that revealed the laws of man as an intelligent and moral being. To obey these laws was to serve the will of the Creator as well as to improve humanity.[144] In the words of Charles Follen (1796–1840): "The great aim of all his inquiries into human nature, was to search out the will of God in the creation of man. Obedience to his laws he considered as the highest wisdom, and most expansive freedom."[145]

Small wonder that Gall complained of Spurzheim that he had "too frequently deviated from the pure path of observation and had thrown himself into ideal-metaphysical and even theological reveries."[146] The fact remains that it was Spurzheim who made phrenology popular in Great Britain and the United States,[147] and it was he who influenced George Combe, the most devoted and most influential of the phrenologists. This remarkable Scottish lawyer, who was born in Edinburgh in 1788, had spent a difficult youth, suffering from illness and overshadowed by the gloom of a strict doctrine of predestination. "Some persons were elected to everlasting enjoyment in heaven; many more passed over by God's decree, before they were born, to everlasting torments in hell. I included myself at once in this category; for the doctrine of Christ's having suffered for *my* sins and purchased my redemption, appeared inconsistent, first with a pre-existing irreversible decree and, secondly, with benevolence and justice." In 1815, Combe was converted to phrenology after he had witnessed Spurzheim dissect a brain. By that time he had become beset with doubts about the orthodox tenets of Calvinism, had studied the works of the Scottish philosophers, and had thought of writing "some useful book on human nature, and especially on the education and intellectual state of the middle ranks of society."[148]

Such a book on human nature Combe eventually did write, the *Essay on the Constitution of Man and its Relations to External Objects*, which first appeared in June 1828 and became the most famous of his many publications. By October 1838, more than seventy thousand copies of this work had been sold in numerous editions in Great Britain as well as the United States.[149]

The book is an application of the doctrine that man has been created in harmony with the world in which he has to live. He is subject to physical and organic laws and endowed with animal, moral, and intellectual powers specified by phrenology. Moral sentiments and intellect, if "illuminated by knowledge of science and of moral and of religious duty,"[150] are supreme over the others, for they enable man to know the laws of nature (including himself) and to live accordingly. The evils afflicting mankind are not inherent in a defective constitution, but due to the infringement of natural laws through ignorance or disobedience. Of course, phrenology as the science of the mind held a key position in this whole structure. But Combe himself put the main emphasis on man's subjection to natural laws rather than on phrenology as a particular doctrine. "We are physical, organic, and moral beings, acting under the sanction of general laws, let the merits of Phrenology be what they may."[151] It was in the relationship of man to these immutable laws that Combe found satisfaction for his religious feelings and the escape from predetermined damnation that he had vainly sought in his youth. In a letter to Dr. Channing, among other things he defined as the proper object of religion

> to embrace within its circuit every interest of man, to direct the mind to the divine origin in the institution of the human being and external nature, to the legitimate use of man's functions as being the true will of God, to every abuse of these as being transgressions of that will; and to God as being in all instances a kind Creator, a benign Father, and an all-wise Ruler of the world, who instituted man that he might enjoy the felicity of his rational nature, and who exacts no service except that obedience which is man's highest advantage to give.[152]

In a delayed answer to this letter, Channing commented upon the success of the *Constitution of Man* in the United States and added that "the common remark, however, is that the book is excellent, in spite of its phrenology."[153] This remark is revealing of the role phrenology was destined to play. Its chief message could be formulated as: man is throughout an object of science and mankind can be improved by spreading scientific knowledge. Consequently, phrenology was a leaven in popularizing science, in furthering educational reform, and in aiming at the liberalization of political and economic conditions. The triumphant welcome ac-

corded to Spurzheim in 1832 and to Combe in 1838, the latter's influence upon Horace Mann, the American educator, and his great success as a lecturer are but examples of the widespread influence of phrenology in the United States, even among those who did not profess to be phrenologists.[154]

In Great Britain, likewise, phrenology reached much further than the group of its outspoken adherents. It became, above all, a movement among what was called "the industrial classes." Opposed alike to the predominance of religious teaching in the schools and the concentration on classical studies, the phrenologists insisted on a widespread secular education in which science and economics should find their due place. Combe had given lectures "to the middle and working classes" from which a "Philosophical Association for procuring Instruction in Useful and Entertaining Science" resulted.[155] Another phrenologist, Mattieu Williams, himself a mechanic, was influential at the London Mechanics' Institution and in 1848 joined forces with Combe and became the principal of the Williams School in Edinburgh, a secular school that drew its pupils from the working classes.[156] Nor could it fail that phrenologists joined in the great reform movement that was then sweeping over England. Combe consistently urged improvements in the conditions of workers and denounced existing political practices of trade restriction, colonial rule, and parliamentary corruption.[157] And not by chance was Richard Cobden, the protagonist in the fight against the corn laws, a friend of Combe and himself a convinced phrenologist. To him, as to many others of his time, phrenology stood for the scientific view of man's nature,[158] and this meant the power of reason against privilege, prejudice, and traditionalism in education, religion, national economy, and politics.

However, English phrenologists tended toward a reformist rather than a revolutionary attitude. Laws had to be changed and democratic institutions perfected, but above all man himself had to be improved as an individual through the exercise of his moral faculties.[159] And phrenology, as Spurzheim had already emphasized,[160] taught tolerance, since it showed that men had been created unequal and could, therefore, not be expected to see everything alike.

Phrenology in England was a forerunner of the much more powerful movement for the spread of scientific education and philosophy that is associated with Thomas Huxley, Tyndall, and others of their generation.

It was not agnostic and materialistic in the sense of many Darwinists; its tenor was rather that of the Bridgewater treatises, which Combe's *Constitution of Man* actually preceded.[161] This may be one of the reasons why the problem of Design became so important in the controversies about Darwinism in England. Nor must it be forgotten that Huxley and Tyndall repeatedly addressed audiences of workingmen who, a generation before, had listened to the phrenologists. It would certainly be a serious mistake to overrate the breadth of the phrenological movement in England. The circle of its serious representatives remained relatively small, and it numbered few physicians and scientists of repute—a serious handicap for a doctrine that claimed to march in the forefront of science. Yet in England, just as in France, a bridge led from phrenology to the science of the later nineteenth century. Herbert Spencer, for instance, in his early years was so deeply interested in phrenology that in 1846 he devised a "cephalograph for taking more exact measurements of the head." Later he became "sceptical about current phrenological views,"[162] but in 1855 he wrote: "Localization of function is the law of all organization whatever: separateness of duty is universally accompanied with separateness of structure: and it would be marvellous were an exception to exist in the cerebral hemispheres."[163] This statement occurs in Spencer's *Principles of Psychology*, a work that exerted such a strong influence upon John Hughlings Jackson.[164] Again, Combe was acquainted with George Eliot, who was a believer in phrenology and who was so closely associated with George Lewes, the positivistic interpreter of Gall.[165] Indeed, phrenology in France and England did not develop in complete isolation; there existed numerous connections, just as there were connections between phrenology and the positivistic physiology of the late decades of the nineteenth century.

Of the phrenological movement the above pages have sketched two aspects, each of which assumed its own momentum, though they were united in the thought of its founder. But as so often happens in history, the work of the man had its own fate; everybody took what he wanted regardless of the intentions of the author. It was just this existence of so many different elements that predisposed phrenology to serve many aims, even beyond those indicated, and attracted persons of divers temperament and outlook.

APPENDIX

Of Gall's *Philosophisch-medicinische Untersuchungun über Natur und Kunst im kranken und gesunden Zustande des Menschen,* only the first volume appeared. It was republished in 1800 and, since I have only had this edition at my disposal, I am unable to say what may have been changed or added after 1791.

The work aims at defining the relationship between nature and its healing power on the one hand and medical art on the other. Repeatedly Gall insists on facts as against speculation, and his contempt for "metaphysicians" is already well pronounced. He presents vast casuistic material from medical and philosophical literature, a document of his erudition but also of his lack of discrimination where so-called experience is concerned. From the outset he attacks the views of Stahl and his followers, who claimed that the soul was active in all vital functions. Instead, Gall maintains that God created all creatures so that through "predestined mechanisms" (p. 45) of their organization there exists harmony between their needs and their abilities to satisfy them. In this respect, humans hold no superiority over animals; they are all equally perfect in themselves (p. 79). It is this predestined order that keeps man, too, in much more rigid limits than his pride and egoism would lead him to believe. "The master plan is outlined for him too. And he stands between barriers which do not seem to impede him, only because his God knew how to combine everything that was to fall into his sphere of action with his happiness and comfort" (p. 118). Yet nature can fail and diseases may arise that warrant the cautious interference of the physician. "The physical world has this in common with the moral world that evil is an inseparable consequence of that institution which is necessary for life, growth, health and preservation" (p. 251).

Where Gall dwells upon the animal nature in humans, he often uses parallels that are in bad taste—for instance, he exemplifies the sphere of sexual functions by a detailed account of the behavior of dogs (p. 28 ff.). If his phrenological lectures were delivered in a similar vein, the disgust felt by some of his listeners would be understandable. The question might even be raised as to whether his cynical tone did not contribute to the accusation of "crass" materialism.

Gall, in the book under discussion, does not deny the existence of the human soul or the possible influence of our notions and imaginings upon our health. But he claims that in normal life the soul is "merely a spectator of the play of the organs; she has all the single ideas and the consequences of ideas not because she wants to have them, but because the movement of the organs compels her to have them" (p. 180). Besides, we "do not yet know of a place in the brain where all nerves join and which, therefore, might be a suitable residence for the soul"

(p. 24). Gall defends the theory "that the various psychic faculties and notions have their seat in different places of the brain" (p. 197 f.). He connects this idea of localization with the unequal endowments of men indicated by the shape of their brains (ibid.) and even with criminal tendencies that occasionally may overpower an individual (p. 677 f.). If these "phrenological" passages actually occur in the first edition, they must be considered highly significant for Gall's early development.

NOTES

The numbers in parentheses refer to the numbered reference list following the notes.

1. Regarding Broussais and Bouillaud see below; for Abernethy see his *Physiological Lectures*, 89 (1). The number of physicians favorably inclined to phrenology was, of course, much greater, and the list of testimonials in favor of phrenology and George Combe, printed in Gibbon's biography (52), 1:318 ff., gives an approximate idea of the status in 1836.

2. See below.

3. See Clapton (24), 260 f., 278 f., and passim, and below.

4. See below.

5. Riegel (101, 102) made an interesting attempt in this direction regarding the United States, and Delaunay (34) did so regarding France.

6. See Macalister (79), Froriep (46), and, in defense of phrenology, Möbius (82) and Hollander (59, 60). Gall's doctrine of isolated psychic faculties and their anatomical localization was reviewed by Riese (103, 104) in two articles to which Dr. E. Ackerknecht kindly drew my attention. At this point, I mention Blondel's book (6), with which the present chapter agrees in many respects.

7. Froriep (46), 38.

8. Gall and Spurzheim (49), 1:ii.

9. Spurzheim (120), 1:10. See also Gall's own testimony in his letter of 1798 to von Retzer, an English translation of which appeared in Capen (22), 70–86.

10. Significantly enough, Lavater (71) chose as the motto of his work (vol. 1, title page): "God created man after His own image." See Delaunay (34), 1242.

11. Gall in Capen (22), 85.

12. One of the first to have sensed materialism and denial of moral liberty in Gall's teachings was the Danish poet W. Schack von Staffeldt (1769–1826), who visited Vienna in 1796; see Neuburger (91), 158, 164, 165. In 1805 the enmity between Gall and the "Naturphilosophen," especially Steffens, was already notorious, as is evidenced by contemporary letters; see Ebstein (38), 292 f. A. W. Schlegel, while denouncing the "crass materialism" of Gall, assigned physiognomy to the sphere of fine arts; see Gode-von Aesch (53), 223n. 23.

13. Möbius (82), 8 f.

14. The first, but very short, outline by Gall himself was his letter to von Retzer, which was published in 1798; see note 9, above.

15. See Capen's biography of Spurzheim (119), 12.

16. Spurzheim (119), xviii.

17. Spurzheim (120), 1:12.

18. Gall and Spurzheim (51).

19. Among these facts may be mentioned the fibrous composition of the white substance of the central nervous system in contrast to the gray masses, which are the organic nutritional matrix of nerve fibers, and the final confirmation of the decussation of the pyramids; see Neuburger (92), 53 and 320, and Ebstein (38), 320.

20. Gall and Spurzheim (51) passim.

21. Gall (48), 2:13 f.

22. Gall and Spurzheim (51), 168.

23. For this phase of Gall's life, see esp. Ebstein (38).

24. Gall and Spurzheim (49), 2:147: "M. le docteur Spurzheim ayant quitté Paris pour enseigner en Angleterre la doctrine des fonctions du cerveau, et pour continuer de recueillir des faits nouveaux, il ne concourt plus à la rédaction de cet ouvrage." Strictly speaking, therefore, Spurzheim's collaboration in the main work extends only over vol. 1 and pp. 1–146 of vol. 2.

25. Spurzheim (120), 1:12.

26. Flourens (43), xiv: "I frequently quote Descartes: I even go further for I dedicate my work to his memory. I am writing in opposition to a bad philosophy, while I am endeavouring to recall a sound one."

27. Capen (22), 72 and 86 (Gall's letter to von Retzer).

28. Gall and Spurzheim (49), 1:xxvii and xxxii.

29. Ibid., 1:107.

30. Blondel (6), 116 ff.; Riese (103), 111. The term *positivistic* is here taken in its general connotation rather than as referring to Comte's philosophy. Where the latter is meant, the word will be capitalized.

31. Riese (103), 109.

32. Gall and Spurzheim (49), 2:3–4.

33. Ibid., 2:4, 2:5.

34. Ibid., 1:106 f.

35. Blondel (6), 50.

36. This list, as well as the following one, is somewhat simplified insofar as Gall gives a wider latitude to each organ, which he defines by more than the one designation chosen by me.

37. Gall and Spurzheim (49), 4:114.

38. Ibid., 115; Blondel (6), 137.

39. Yet even these faculties are not entirely restricted to human beings. Goodness, e.g., can also be observed in animals; see Gall and Spurzheim (49), 4:157, Lélut (73), 40.

40. Blondel (6), 57, 82.

41. Gall and Spurzheim (49), 4:240.

42. Ibid., 4:126.

43. Gall and Spurzheim (49), 2:69: "La liberté morale n'est donc autre chose que la faculté d'être déterminé et de se déterminer par des motifs." For this Gall quotes in a footnote: "Bonnet, Palingénésie, T. I, p. 27." I have not been able to find this sentence literally in Charles Bonnet's *La palingénésie philosophique* (9) or his *Essai analytique* (8), although the whole line of thought is very similar. Gall's acquaintance with Bonnet is, of course, well known, and the wide range of this relationship was emphasized by Bentley (4), 110. At this point I stress that the present chapter is not concerned with Gall's relative originality and his sources.

44. Gall and Spurzheim (49), 2:73.

45. Ibid.

46. Flourens (43), 42.

47. Lange (68), 2:338.

48. Gall and Spurzheim (49), 2:74.

49. Ibid., 2:75.

50. Ibid.

51. Ibid., 1:iii.

52. Ibid., 2:29.

53. Ibid., 1:iii–iv.

54. Ibid., 2:16, 2:30 ff. Cf. Soury (112), 510.

55. Gall and Spurzheim (49), 2:4, 2:34 f. In outlining Gall's "physiological theory of knowledge," Blondel (6), 154, rightly stresses the analogy with Kant. "Peut-être sommes-nous autorisé à y voir une sorte d'idéalisme organologique, une rudimentaire transposition du système de Kant, où ce sont des organes qui tiennent l'emploi de formes de la sensibilité et de catégories de l'entendement." Cf. also Gall and Spurzheim (49), 1:107, which reads as if Gall, moreover, wished to mediate between German idealism and French sensualism, and Lélut (72), 360.

56. Gall and Spurzheim (49), 2:81, 2:95.

57. Neuburger (90), 60n. 22.

58. Gall and Spurzheim (49), 2:100 ff., 2:105 ff., where he even goes so far as to recommend slow and painful capital punishment in extreme cases.

59. Ibid., 2:98.

60. Ibid., 2:98–146; Neuburger (90), 60n. 22.

61. For the following see Temkin (125).

62. Cabanis (21), 81 ff.

63. Cabanis (21), 305 f., recognizes centers of sensibility apart from the brain and spinal cord, and on 501n. he clearly admits a "sensibilité sans sensation," which, however, is supposed to differ from mere irritability.

64. Temkin (124).

65. In the philosophical literature this tradition apparently goes back to Damiron (33), although he was doubtful of making Gall a mere sensualist. Dilthey (37), 534–35, mentions the following as the main representatives of the French sensualist school: "Condillac, Cabanis, … Destutt de Tracy, … Volney, und die Vertreter der physiologischen Begründung der Psychologie, Broussais und Gall." I owe this passage to the courtesy of Dr. George Rosen.

66. Gall and Spurzheim (49), 4:227.

67. Ibid., 1:58, 1:60 f.

68. Flourens (43), 27 f., and (41), 156 ff., 160 ff.

69. Gall and Spurzheim (49), 1:107. The passage "l'éducation dont le but doit être de faire ce que l'on désire des individus et des nations" is particularly interesting. Such an idea of the aim of education harmonized very well with the concept of education as a process of "conditioning."

70. Ibid., 4:227 f. Blondel (6), 83 ff., and Delaunay (34), 1241 ff., rightly stressed Gall's opposition to the sensualists. It must be added that this opposition extended to their social and political ideas as well.

71. Blondel (6), 117 f.

72. Temkin (125), 13.

73. Lélut (72), 360 ff., pointed out the parallels between the Scottish school and phrenology. Combe (25), in the preface to his *Constitution of Man,* referred to Hutcheson, Adam Smith, Reid, Stewart, and Thomas Brown. In this connection I regret that the book by Garnier, *De la phrénologie et de la psychologie comparées,* 1839, which is quoted by Lévy-Bruhl (74) 206, was not available to me, nor could I consult F. Dubois, *Examen des doctrines de Cabanis, Gall, et Broussais* (Paris, 1842), which is mentioned by Peisse in Cabanis (21), lvi, footnote.

74. Boas (7), 251.

75. Ibid., 119.

76. Picavet (96), 477.

77. Broussais (16), 490.

78. Ibid., 2–3.

79. Ibid., 543n.

80. Saucerotte (108), col. 535.

81. Peisse in Cabanis (21), 62n.

82. Broussais (16), xii.

83. Broussais (18), 2:389: "M. le professeur Richerand se range du côté de Cabanis pour rapporter aux viscères les déterminations instinctives; et la vérité de ce fait ne parait plus aujourd'hui contestée que par m. le docteur Gall."

84. Broussais (16), 426.

85. Ibid., 478.

86. Ibid., 479.

87. Ibid., 468 ff. A somewhat similar appreciation already occurs in Broussais (19), 68 ff.

88. Broussais (15), 2, 86.

89. This has to be pointed out in view of the accusation that Broussais had become a phrenologist to give new impetus to his waning popularity; see Huet (61), 44 f. Casimir Broussais (17), 1:xiii, was quite right when he stressed the unity of his father's work, including the phrenological phase. Similarly, Lélut (73), 225 f., although he was highly critical of phrenology.

90. Comte (31), 4:iii–iv. My attention was drawn to this review by Hayek's article (55).

91. Ibid., 598. Possibly here, as in his general biological views, Comte had been influenced by de Blainville. The latter in 1821 was reported as upholding phrenology [see Capen in Spurzheim (119), 59 f., and Delaunay (34), 1279], although in his *Cours* (5), 2:369, 2:408, he expressed himself very cautiously.

92. Comte (31), 4:645, 4:647, 4:648.

93. Comte (28), 3:541, 3:550.

94. In this connection see particularly Boas (7) and Lévy-Bruhl (74), 188 ff.

95. Comte (28), 3:530, 3:567, 3:585 ff., 3:586–87, 3:567 ff.

96. Comte (31), 4:650; see also Lewes (75), 407 f.

97. Huxley (63), 149, 154 f.

98. Boas (7), 284; cf. Hayek (55).

99. Flourens (43), xiii: "The seventeenth century recovered from the philosophy of Descartes; the eighteenth recovered from that of Locke and Condillac: is the nineteenth to recover from that of Gall?"

100. In his controversy against phrenology, Flourens appears as the physiologist of the "eclectic" school, a possibility that I intend to investigate on a broader basis.

101. Ferraz (40), 308.

102. Lewes (75), 421 ff.; Littré (78), 541 f.

103. Lewes (75), 408; Littré (78), 542. See also, among moderns, Boring (10), 55.

104. Littré (78), 545.

105. Lewes (75), 421. For Positivism as a transmitter of phrenological ideas, see also Delaunay (34), 1284.

106. I am, of course, aware that the distinction between positivism and metaphysics is a matter of convenience rather than of substance. In particular, I agree with Riese (103), 121 f., that Gall's "faculties" have a metaphysical basis.

107. Neuburger (90), 65, 35, 30; Neuburger (89), 100.

108. Gall and Spurzheim (49), 2:60 f.

109. Neuburger (90), 35.

110. Gall and Spurzheim (49), 4:190 f.

111. See the article by Spector (113).

112. See, e.g., Gall and Spurzheim (49), 4:259 f. Gall (47), 130, quotes Galen with approval.

113. Gall and Spurzheim (49), 1:iv.

114. See above, Appendix.

115. Gall and Spurzheim (49), 2:37. Cf. Blondel (6), 64 f.

116. Gall (48), 1:173–75 (contracted).

117. Gall and Spurzheim (49), 4:251 f., 4:252.

118. Ibid., 4:252. Cf. Soury (112), 510.

119. Gall and Spurzheim (49), 4:253.

120. Ibid., 4:254, 4:255, 4:256.

121. On November 18, 1805, in the midst of the Napoleonic wars, Gall wrote: "Ach wie vieler Menschen Glück stört das unselige Kriegen ! und doch liegts, wie es meine Lehre zeigt, im Plane der Schöpfung." Neuburger (90), 14; cf. ibid., 59n. 10.

122. Gall and Spurzheim (49), 2:65. Cf. above, Appendix.

123. Gall and Spurzheim (49), 4:247, 4:260.

124. Ibid., 2:66: "C'est ainsi que, pour le mal moral, le genre humain porte sur le front l'empreinte du mal originel."

125. Comte (31), 4:648.

126. In the face of Comte's attitude, Blondel (6), 125, 153, rightly insists on the sincerity of Gall's belief in God.

127. Broussais (15), 340.

128. Maistre (80), 1:46 ff.

129. Cf. Flourens (41), 124, where he also refers to Chateaubriand; it is characteristic that the latter had no use for Gall; see Chateaubriand (23), 2:302 f. see also Delaunay (34), 1284.

130. Damiron (33), 1:190n. According to Delaunay (34), 1279, the French Société phrénologique was founded in 1831 under liberal auspices.

131. Marie (81), 570. Cf. Riese (103), 110, and Rolleston (107), 1260.

132. Neuburger (90), 37.

133. Ibid., 63n. 48.

134. Cf. Fossati (45), col. 280.

135. Gall (48), 2:50; cf. Soury (112), 511.

136. Gall and Spurzheim (49), 4:248.

137. Speaking of the criminal whose acts correspond with his inclinations, Gall and Spurzheim (44), 2:102, write "Ce côté de l'homme dépravé pourra bien déplaire à plusieurs de ces hommes qui ne rêvent que les grandeurs de l'espèce humaine." Similarly, ibid., 100, the authors say that mere punishment does not improve malefactors but add, "D'un autre côté, de prétendus philosophes ont imaginé toutes sortes de rêveries sur les droits de l'homme," etc.

138. Ibid., 2:32: "Ces exemples prouvent encore la justesse du principe qu' Aristote . . . . a posé comme formant la base de toute la science politique; savoir que, dans ce monde, les uns sont nés pour dominer, et les autres pour obéir."

139. Post (97), esp. 229.

140. Barclay (2), 380 f.

141. Spurzheim (118), 123, 315.

142. Ibid., 340 f. Delaunay (34), 1248 f., quite unnecessarily attributes this change to Spurzheim's "prudence."

143. Spurzheim (118), 356 f.

144. Cf. Spurzheim's preface to his "Catechism," as reprinted in Capen (22), 241.

145. Capen's biography of Spurzheim (119), 145.

146. Neuburger (90), 44 f.

147. Lélut (73), 217. According to Fossati (45), col. 281, Gall's journey to London in 1823 proved disappointing.

148. Gibbon (52), 1:39, 1:92, 1:73.

149. Ibid., 1:262n.

150. Combe (25), chap. 2, sec. 4, p. 10.

151. Ibid., preface to the edition of 1828.

152. Gibbon (52), 1:219 f.

153. Ibid., 221.

154. See the account of Combe's American tour as described by Gibbon (52); Riegel (101, 102).

155. Gibbon (52), 1:254, 1:300, 1:289.

156. Ibid., 2:247 ff. Although Williams remained a phrenologist to the end, it is interesting to see how with him the cranioscopic aspect became of secondary importance. In his *Vindication of Phrenology* (127), 256 f., he wrote: "No phre-

nologist properly so called pretends to read, or delineate, or determine character by a simple examination of the head. He knows that the proportionate development of different regions of the brain is but one of the factors of character, the other being the total of the external influences that have operated on the
individual. These include his life-history, his formal and his informal education;
what he has been deliberately taught by parents and other avowed teachers, and
the still more potent educational influences of the incidents of daily life."

157. Combe (25), chap. 3, sec. 3.

158. Gibbon (52), passim, and Morley (86), 26, 60, 78, 128.

159. Combe (25), 54; Gibbon (52), 1:302.

160. Spurzheim (118), 359 f.

161. See Gibbon (52), 296, where Combe is mentioned as complaining of
Thomas Chalmers who, in his Bridgewater treatise *On the Adaptation of External
Nature to the Moral and Intellectual Constitution of Man*, adopted the principles of the
Constitution of Man without referring to it.

162. Spencer (114), vol. 1, appendix H; 1:540.

163. Spencer (115), 606 f.

164. Temkin (123), 301–5.

165. See Cross (32), 78. In 1855 Eliot wrote (279), "We are reading Gall's
'Anatomie et Physiologie du Cerveau,' and Carpenter's 'Comparative Physiology,'
aloud in the evenings"—another remarkable example of the close association of
phrenology and orthodox physiology among laymen.

REFERENCES

1. Abernethy, John. *Physiological Lectures,* 2nd ed. London, 1822.

2. Barclay, John. *An Inquiry into the Opinions, Ancient and Modern, Concerning Life and
   Organization.* Edinburgh, 1822.

3. Benedikt, Moriz. *Anatomical Studies upon Brains of Criminals,* trans. from the
   German by E. P. Fowler. New York, 1881.

4. Bentley, Madison. "The psychological antecedents of phrenology." *Psychol
   Monogr* 21 (1916, 4, no. 92): 102–15.

5. Blainville, Ducrotay de. *Cours de physiologie générale et comparée,* 2 vols. Paris,
   1829.

6. Blondel, Charles. *La psycho-physiologie de Gall.* Paris: Alcan, 1914.

7. Boas, G. *French Philosophies of the Romantic Period.* Baltimore: Johns Hopkins
   Press, 1925.

8. Bonnet, Charles. *Essai analytique sur les facultés de l'âme.* Neuchatel: Samuel
   Fauche, 1782 [Oeuvres d'histoire naturelle et de philosophie de Charles
   Bonnet, vol. 6].

9. ———. *La palingénésie philosophique.* Neuchatel: Samuel Fauche, 1783 [Oeuvres d'histoire naturelle et de philosophie de Charles Bonnet, vol. 7].

10. Boring, Edwin G. *A History of Experimental Psychology.* New York: D. Appleton-Century Co., 1929.

11. ———. *Sensation and Perception in the History of Experimental Psychology.* New York: D. Appleton-Century Co., 1942.

12. Bouillaud, J. *Essai sur la philosophie médicale et sur les généralités de la clinique médicale.* Paris, 1836.

13. ———. "Recherches cliniques propres à démontrer que la perte de la parole correspond à la lésion des lobules antérieurs du cerveau, et à confirmer l'opinion de M. Gall, sur le siège de l'organe du langage articulé." *Arch Gen Med* 8 (1825): 25–45.

14. Bromberg, Walter. "Some social aspects of the history of psychiatry." *Bull Hist Med* 11 (1942): 117–32.

15. Broussais, F.-J.-V. *Cours de phrénologie.* Paris, 1836.

16. ———. *De l'irritation et de la folie.* Paris, 1828.

17. ———. *De l'irritation et de la folie,* 2nd ed., 2 vols. Paris: Casimir Broussais, 1839.

18. ———. *Examen des doctrines médicales et des systèmes de nosologie,* 2 vols. Paris, 1821.

19. ———. *A Treatise on Physiology Applied to Pathology,* trans. J. Bell and R. La Roche. Philadelphia, 1826.

20. Burdach, Karl Friedrich. *Vom Baue und Leben des Gehirns,* 3 vols. Leipzig, 1819–26.

21. Cabanis, P.-J.-G. *Rapports du physique et du moral de l'homme,* 8th ed. with notes by L. Peisse. Paris, 1844.

22. Capen, Nahum. *Reminiscences of Dr. Spurzheim and George Combe.* New York, 1881.

23. Chateaubriand, F. A. R. *Mémoires d'outre-tombe,* new ed. by E. Biré, 6 vols. Paris, [1898–1900].

24. Clapton, G. T. "Lavater, Gall et Baudelaire." *Rev Littérature Comparée* 13 (1933): 259–98, 429–56.

25. Combe, George. *The Constitution of Man.* New York, 1835.

26. ———. *The Life and Correspondence of Andrew Combe, M.D.* Philadelphia, 1850.

27. ———. *A System of Phrenology,* 3d ed. Edinburgh, 1830.

28. Comte, Auguste. *Cours de philosophie positive,* 2nd ed. by É. Littré, 6 vols. Paris: Baillière, 1864.

29. ———. *Early Essays on Social Philosophy,* trans. Henry Dix Hutton, new ed. by Frederic Harrison (New Universal Library).

30. ————. *The Positive Philosophy of Auguste Comte*, trans. and condensed by Harriet Martineau. New York, 1855.

31. ————. *System of Positive Polity*, 4 vols. London: Longmans, Green and Co., 1877.

32. Cross, J. W., *George Eliot's Life as Related in Her Letters and Journals*, 3 vols. New York: Harper Brothers, 1885, vol. 1.

33. Damiron, Ph. *Essai sur l'histoire de la philosophie en France, au XIXe siècle*, 3d ed., 2 vols. Paris, 1834.

34. Delaunay, Paul. "De la physiognomonie à la phrénologie." *Progrès méd.* 29–31 (1928): 1207–11, 1237–51, 1279–90.

35. Demangeon, J. B. *Physiologie intellectuelle ou développement de la doctrine du Professeur Gall sur le cerveau et ses fonctions.* Paris, 1806.

36. Dessoir, Max. *Abriss einer Geschichte der Psychologie.* Heidelberg, 1911 [Die Psychologie in Einzeldarstellungen, vol. 4].

37. Dilthey, Wilhelm. "Die drei Grundformen der Systeme in der ersten Hälfte des 19 Jahrhunderts." *Gesammelte Schriften* (1925): 528–54.

38. Ebstein, Erich. "Franz Joseph Gall im Kampf um seine Lehre." In *Essays on the History of Medicine Presented to Karl Sudhoff*, ed. Charles Singer and Henry E. Sigerist. London, 1924, 269–322.

39. ————. "Gall in defense of his theory." *Med Life* 30 (1923): 369–72.

40. Ferraz, M. *Étude sur la philosophie en France au XIXe siècle.* Paris, 1877.

41. Flourens, P. *De la phrénologie et des études vraies sur le cerveau.* Paris, 1863.

42. ————. *Examen de la phrénologie.* Paris, 1842.

43. ————. *Phrenology Examined*, trans. from 2nd ed. of 1845 by Charles De Lucena Meigs. Philadelphia, 1846.

44. ————. "Recherches physiques sur les propriétés." *Arch Gen Med* 2 (1823): 321–70.

45. Fossati. "Gall, François-Joseph." *Nouvelle Biographie Générale* 19 (1857): cols. 271–84.

46. Froriep, August. *Die Lehren Franz Joseph Galls beurteilt nach dem Stand der heutigen Kenntnisse.* Leipzig: Barth, 1911.

47. Gall, F. Joseph. *Philosophisch-medicinische Untersuchungen über Natur and Kunst im kranken und gesunden Zustande des Menschen*, 2nd ed., vol. 1. Leipzig, 1800.

48. ————. *Sur les fonctions du cerveau et sur celles de chacune de ses parties*, 6 vols. Paris, 1822–25.

49. Gall, F. J., and G. Spurzheim. *Anatomie et physiologie du système nerveux en général, et du cerveau en particulier, avec des observations sur la possibilité de reconnoitre plusieurs dispositions intellectuelles et morales de l'homme et des animaux, par la configuration de leurs têtes*, 4 vols. and atlas. Paris, 1810–19 (vols. 3 and 4 have Gall as the sole author).

50. ————. *Des dispositions inées de l'ame et de l'esprit, du matérialisme, du fatalisme et de la liberté morale, avec des réflexions sur l'éducation et sur la législation criminelle.* Paris, 1811.

51. ————. *Recherches sur le système nerveux en général, et sur celui du cerveau en particulier; Mémoire présenté à l'Institut de France, le 14 Mars 1808; suivi d'observations sur le rapport qui en a été fait à cette compagnie par ses commissaires.* Paris, 1809.

52. Gibbon, Charles. *The Life of George Combe,* 2 vols. London, 1878.

53. Gode-von Aesch, Alexander. *Natural Science in German Romanticism.* New York: Columbia University Press, 1941.

54. Gouhier, Henri. *La vie d'Auguste Comte.* Paris: Librairie Gallimard, 1931 (Vies des Hommes Illustres, No. 63).

55. Hayek, F. A. v. "The counter-revolution of science." *Economica,* new ser. 8 (1941): 9–36, 119–50, 281–320.

56. ————. "Scientism and the Study of Society." *Economica,* new ser., 9 (1942): 267–91; 10 (1943): 34–63; 11 (1944): 27–39.

57. Holland, Henry. *Chapters on Mental Physiology,* 2nd ed. London, 1858.

58. Hollander, Bernard. "The centenary of Francis Joseph Gall, 1758–1828." *Med Life* 35 (1928): 373–80.

59. ————. *In Search of the Soul,* 2 vols. New York: Dutton, n.d.

60. ————. *The Unknown Life and Works of Dr. Francis Joseph Gall.* London, 1909.

61. Huet, Jeanne. *Broussais, sa vie et son oeuvre (1772–1838).* Diss., Paris, 1938.

62. Huschke, Emil. *Schaedel, Hirn und Seele des Menschen und der Thiere nach Alter, Geschlecht und Raçe.* Jena, 1854.

63. Huxley, Thomas Henry. "The scientific aspects of positivism [1869]." In *Lay Sermons, Addresses, and Reviews.* New York: Appleton, 1878, 147–73.

64. James, William. *The Principles of Psychology,* 2 vols. New York: Henry Holt and Co., 1923.

65. Kitchel, Anna Theresa. *George Lewes and George Eliot: A Review of Records.* New York: John Day Co. [1933].

66. Klinckowstroem, Graf Carl v. "Zur Gall-Biographie." *Mitteilungen Geschichte Med Naturwissenschaften* 20 (1921): 286–87.

67. Ladd, George T. *Elements of Physiological Psychology.* New York: Scribner's, 1887.

68. Lange, Friedrich Albert. *Geschichte des Materialismus und Kritik seiner Bedeutung in der Gegenwart,* 2nd ed. Buch, Iserlohn: J. Baedeker, 1875.

69. Lashley, K. S. *Brain Mechanisms and Intelligence.* Chicago: Univ. Chicago Press, 1929.

70. ————. "Integrative functions of the cerebral cortex." *Physiol Rev* 13 (1933): 1–42.

71. Lavater, Johann Caspar. *Physiognomische Fragmente, zur Beförderung der Menschenkenntniss und Menschenliebe.* Leipzig, 1775 (reproduced 1908).

72. Lélut, F. *Qu'est-ce que la phrénologie?* Paris, 1836.

73. ————. *Rejet de l'organologie phrénologique de Gall, et de ses successeurs.* Paris, 1843.

74. Lévy-Bruhl, L. *The Philosophy of Auguste Comte*, authorized translation. London: Swan Sonnenschein and Co., 1903.

75. Lewes, George Henry. *The History of Philosophy from Thales to Comte*, 3d ed. [in 2 vols.], vol. 2. London: Longmans, Green and Co., 1867.

76. ————. "The modern metaphysics and moral philosophy of France." *Brit Foreign Rev* 15 (1843): 353–406.

77. Liepmann, H. "Franz Joseph Gall." *Deutsche Med Wochenschrift* 35 (1909, no. 1): 979–80.

78. Littré, E. *Auguste Comte et la philosophie positive*, 2nd ed. Paris, 1864.

79. Macalister, Alexander. "Phrenology." In *Encyclopaedia Britannica*, 11th ed. 1911, 21:534–41.

80. Maistre, T. de. *Les soirées de Saint-Pétersbourg*, 7th ed., 2 vols. Lyon, 1854.

81. Marie, Pierre. "Revision de la question de l'aphasie: l'aphasie de 1861 à 1866; essai de critique historique sur la genèse de la doctrine de Broca." *Sem Med* 26 (1906): 565–71.

82. Möbius, P. J. *Franz Joseph Gall.* Leipzig: Barth, 1905.

83. Morin, Georges. "Gall et Goethe: Goethe disciple de Gall." *Paris Med*, Partie paramédicale, 72 (1929): 425–32.

84. ————. "Les ennemis de la phrénologie." *Paris Med*, Partie paramédicale, 72 (1929): 163–69.

85. ————. "Le système de Gall et la psycho-physiologie." *Paris Med*, Partie paramédicale, 72 (1929): 83–87.

86. Morley, John. *The Life of Richard Cobden*, new ed. London: Chapman and Hall, 1883.

87. Mourgue, Raoul. *La philosophie biologique d'Auguste Comte.* Lyon, 1909 [Extrait des Archives d'Anthropologie criminelle et de Médecine légale, October–December 1909].

88. [Nacquart, J. B.]. *Traité sur la nouvelle physiologie du cerveau.* Paris, 1808.

89. Neuburger, Max. "Anhang zu den Briefen Galls." *Arch Geschichte Med* 11 (1919): 93–101.

90. ————. "Briefe Galls an Andreas und Nannette Streicher." *Arch Geschichte Med* 10 (1917): 3–70.

91. ————. *Das alte medizinische Wien in zeitgenössischen Schilderungen.* Vienna, 1921.

92. ————. *Die historische Entwicklung der experimentellen Gehirn-und Rückenmarks-physiologie vor Flourens.* Stuttgart: Enke, 1897.

93. ————. "Zur Gall-Biographie." *Mitteilungen Geschichte Med Naturwissenschaften* 18 (1919): 159.

94. Nivelet, F. *Gall et sa doctrine.* Paris, 1890.

95. Ockenden, R. E. "George Henry Lewes (1817–1878)." *Isis* 32 (1947, no. 85): 70–86.

96. Picavet, François. *Les idéologues.* Paris, 1891.

97. Post, Albert. *Popular Freethought in America, 1825–1850.* New York: Columbia Univ. Press, 1943.

98. Reis, Paul. *Étude sur Broussais et sur son oeuvre.* Paris, 1869.

99. Ribéry, C. "La phrénologie en Amérique." *Rev Philos France Étranger* 55 (January–June 1903): 176–86.

100. Richerand, A. *Elements of Physiology,* trans. from French by G. J. M. De Lys, with notes by N. Chapman, 5th London ed. Philadelphia, 1813.

101. Riegel, Robert E. "Early phrenology in the United States." *Med Life* 37 (1930): 361–76.

102. ———. "The introduction of phrenology to the United States." *Am Hist Rev* 39 (October 1933): 73–78.

103. Riese, Walther. "F.-J. Gall et le problème des localisations cérébrales." *Hygiène Mentale* 31 (1936): 105–36.

104. ———. "Les discussions du problème des localisations cérébrales dans les sociétés savantes du XIXᵉ siècle et leurs rapports avec des vues contemporaines." *Hygiène Mentale* 31 (1936): 137–58.

105. [Robinson, Henry Crabb]. *Some Account of Dr. Gall's New Theory of Physiognomy, . . . with the Critical Strictures of C. W. Hufeland.* London, 1807.

106. Rolleston, J. D. "F. J. V. Broussais (1772–1838): his life and doctrines." *Proc R Soc Med* 32 (1939): 405–13.

107. ———. "Jean Baptiste Bouillaud (1796–1881): a pioneer in cardiology and neurology." *Proc R Soc Med* 24 (1931): 1253–62.

108. Saucerotte, C. "Broussais." In *Nouvelle biographie générale,* vol. 7. Paris, 1854, col. 531–36.

109. Schneider, K. C. "Franz Joseph Gall." *Wien Klin Rundschau* 20:687–690.

110. Schulz, Hans. "Gleichzeitige Stimmen über Galls Vorlesungen in Halle." *Arch Geschichte Med* 12 (1920): 59–69.

111. Selling, Lowell S. *Men against Madness.* New York: Greenberg, 1940.

112. Soury, Jules. *Le système nerveux central.* Paris, 1899.

113. Spector, Benjamin. "Sir Charles Bell and the Bridgewater treatises." *Bull Hist Med* 12 (1942): 314–22.

114. Spencer, Herbert. *An Autobiography,* 2 vols. London: Williams and Norgate, 1904.

115. ———. *The Principles of Psychology.* London, 1855.

116. Spurzheim, J. G. *The Anatomy of the Brain, with a General View of the Nervous System,* trans. from the unpublished French Ms. by R. Willis, 1st Am. ed., rev. Charles H. Stedman. Boston: Marsh, Capen and Lyon, 1834.

117. ———. *Examination of the Objections Made in Britain against the Doctrines of Gall and Spurzheim*. Edinburgh, 1817.

118. ———. *Observations sur la phraenologie*. Paris: Treuttel et Würtz, 1818.

119. ———. *Phrenology, in Connexion with the Study of Physiognomy*, 1st Am. ed., with a biography of the author by Nahum Capen. Boston, 1833.

120. ———. *Phrenology, or the Doctrine of the Mental Phenomena*, 4th Am. ed., 2 vols. Boston, 1835.

121. Stephen, Leslie. "George Combe." *Dictionary of National Biography* 4 (1908): 883–85.

122. Tausig, Paul. "Briefe von Andreas und Nanette Streicher an Anton Franz Rollett über die Gallsche Schädelsammlung." *Arch Geschichte Med* 12 (1920): 50–58.

123. Temkin, Owsei. *The Falling Sickness*. Baltimore: Johns Hopkins Press, 1945.

124. ———. "Materialism in French and German physiology of the early nineteenth century." *Bull Hist Med* 20 (1946): 322–27.

125. ———. "The philosophical background of Magendie's physiology." *Bull Hist Med* 20 (1946): 10–35.

126. Wilks, Samuel. "Notes on the history of the physiology of the nervous system, taken more especially from writers on phrenology." *Guy's Hosp Rep* 3d ser., 24 (1879): 57–94.

127. Williams, W. Mattieu. *A Vindication of Phrenology*. London: Chatto and Windus, 1894.

128. Wundt, Wilhelm. *Grundzüge der physiologischen Psychologie*. Leipzig: Engelmann, 1874.

CHAPTER 7

Historical Reflections on the Scientist's Virtue

It is indeed a privilege to give a lecture in honor of George Sarton, who was so passionately dedicated to the history of science. His passion was an emotional response to the broad meaning that he gave to the subject. He thought of it as "the leading thread in the history of civilization" because science was the most powerful civilizing factor in the evolution of man, "the most precious patrimony of mankind."[1] The cultivation of disinterested science was one of the best ways of purifying humanity, and disinterested science was "the love of truth—as a scientist loves it, the whole of it, pleasant or unpleasant, useful or not; the love of truth, not the fear of it; the hatred of superstition, no matter how beautiful its disguises may be."[2]

If we take "virtue" to mean that which gives a profession its particular moral worth, dedication to disinterested truth was Sarton's notion of the scientist's virtue. And indeed, this seems to agree with the widespread acceptance of science as the pursuit of truth. Whatever moral obligations science may have—and they are vigorously discussed in our days: its role in war and peace, the solution of technical problems, aid in the solution of social problems—science itself rests upon the scientist's willingness to search for the truth. A scientist who falsifies or suppresses data, who intentionally shuns verification of results, is a bad scientist, not merely a mediocre one, and not merely a cheat.

The scientist's relation to truth is the theme on which I offer some historical variations. The metaphysical nature of truth is for philosophers to

Originally published in *Isis* 60, pt. 4, no. 204 (1969). © The University of Chicago Press.

argue. I wish to see what historical roots can be found for Sarton's belief and for his enthusiasm, even at the peril of occasional disagreement with him.

Indeed, Plato, whom, I am afraid, Sarton did not like,[3] reminds us not to substitute a man's authority for objective examination. When Socrates questioned Phaidros about the nature of the soul, Phaidros in replying referred to the great Hippocrates, only to be told that though Hippocrates was right, we should not be satisfied with citing his name. "See what Hippocrates *and* truthful argument (*ho alethēs logos*) say about nature."[4]

So admonished, we add truthful argument to Sarton's authority. There are several passages in Plato where he insists that no man, not even Homer, "ought to be humored in preference to truth."[5] Similar sentiments were expressed by Aristotle, who had criticized his own teacher, Plato. "Still perhaps it would appear desirable, and indeed it would seem obligatory, especially for a philosopher, to sacrifice even one's closest personal ties in defense of the truth. Both are dear to us, yet 'tis our duty to prefer the truth."[6] Out of statements like this grew the anonymous, much quoted saying: "Amicus Plato, sed magis amica veritas" [Plato is my friend, but truth is more so].[7]

There is something in the above quotations from Plato and Aristotle that warns us not to read them in the spirit of the seventeenth century, when the new philosophy conquered the old scholastic ways. Plato and Aristotle do not speak about authority in the abstract. Aristotle does not deny his bond of friendship with Plato; he only claims that the bond with truth must have precedence. If necessary, truth demands the sacrifice of friendship.

This note of sacrifice for truth finds many expressions in antiquity, and I shall mention but one, which goes back to Galen, the great scientist-physician of the second century, whose influence remained powerful for centuries to come. In a very short treatise, Galen defended the thesis that the ideal physician must needs be a philosopher. He must know anatomy and the genera and species of disease, he must despise riches and love work, and he must resist his passions. "It is established that the true physician is a friend of moderation, as well as a disciple of truth," nor will he dispute over mere words; "rather he will explore the truth of the things themselves."[8] As an anatomist, he must also overcome social prejudices. Galen had left the menial task of skinning dead monkeys to slaves, as his pred-

ecessors had done, because it was beneath his dignity. But when he found an unattached piece of muscle, he began to skin the animals himself.[9]

The modern notion of a scientist did not yet exist in Galen's time, nor did it exist at the time of William Harvey, who was an anatomist and a physician. He was a member of the *republic of letters*—I understand that this term first came into use around that time[10]—which united philosophers, humanists, and writers of different nationalities and religions. "I avow myself the partisan of truth alone," wrote Harvey in the dedication of his *De motu cordis* of 1628, and George Ent, in the dedicatory letter to Harvey's *De generatione animalium*, told the College of Physicians that with this book Harvey presented "an offering to the benefit of the republic of letters (*reipublicae literariae*), to your honour, to his own eternal fame."[11] Galileo, who was Harvey's contemporary, referred to himself as a philosopher, a mathematician, and a member of one of the earliest scientific societies, the Accademia dei Lincei, which, in practice, limited itself to experimental work.[12]

But experimental philosophy, though defeating scholastic philosophy and attractive to a new type of man, was not yet identical with science. It only meant that, within the new breed of "philosophers," men whom in retrospect we call *scientists* searched for truth by methods we call scientific. Dedication to truth is not defined by method, though method may indicate what kind of truth is looked for.

What comes to our mind is the association of science with usefulness, or at least with potential usefulness. That knowledge will be useful and that material necessity may have led men to the invention of arts and then to the cultivation of knowledge for its own sake is an idea at least as old as Aristotle.[13] But that philosophy of a new kind should be pursued because of its usefulness is a postulate that gained strength only with Francis Bacon and Descartes. In his *Method for the Easy Understanding of History* of 1566, Jean Bodin looked upon history as the master of life. Human history was useful because it taught man prudence. Natural history—we may cautiously say science—dealt with the inevitable, and its virtue was knowledge that distinguished true from false.[14]

If, later on, science was to claim that in the final analysis all true knowledge was scientific knowledge, this was a heritage from its ancestor the philosopher, not from its other ancestor, the useful craftsman. The Hippocratic Oath, a moral inspiration for the craft of medicine even before

it considered itself a science, went as far as to demand of the physician that he live his life "in purity and holiness." But it did not explicitly bind him to truth.

Speaking generally, the sacrifices that the philosopher was to make for the disinterested pursuit of truth were not necessarily asked of the craftsman. "The laborer is worthy of his reward," says I Timothy 5:18, and so says the scientist of himself, pointing to his social usefulness and demanding material rewards from the society he serves. But St. John 8:32 reads: "And you shall know the truth and the truth shall make you free." I am afraid that the two texts are not easily combined in the same sermon.

But I may not be allowed to let the matter rest there. Neither philosophy nor dedication are to the point, it may be argued. What is involved is just curiosity—the often cited curiosity of the scientist—and this, in turn, is nothing but specialized human curiosity. Did not Aristotle open his *Metaphysics* with the statement that "all men, by nature, desire to know"? And did not Thomas Hobbes in the *Leviathan* define curiosity as the "Desire to know why, and how"? Indeed, his making curiosity a passion characteristic of man and "a lust of the mind, that by a perseverance of delight in the continued and indefatigable generation of knowledge, exceedeth the short vehemence of any carnal pleasure,"[15] seems to devaluate the merit of the sacrifices that the philosopher was willing to take upon himself in the pursuit of truth.

Hobbes's phrasing invites a psychological study of the formation of scientific curiosity, for which I refer to Prof. Lewis Feuer's book.[16] I shall be satisfied with the acknowledgment that curiosity finds pleasure in its satisfaction, and I shall not ask whether this pleasure has biological roots or whether, as William James thought, it differs from the curiosity that animals show.[17] Indeed, if we admit that "man . . . is not only an incurably inquisitive but an incurably ratiocinative being, and [that] the exercise of this function, as of others, carries its own pleasure with it," then we must admit that philosophical and scientific activity will look for an outlet, and not only when called upon to serve other ends. This, I think, was Arthur Lovejoy's meaning in the passage from which I just quoted. He had in mind the recognition of "a nice distinction," discovery of "a new truth," the feeling of "reasoning well and coercively," activities that include the philosopher, the historian, and the scientist.[18]

The element of pleasure in giving way to one's curiosity carries a seri-

ous consequence. The scientist who indulges in it can be accused of self-ishness; indeed, he may even accuse himself and warn his colleagues against taking up residence in the ivory tower. The suspicion of merely following one's own desires, disregarding useful occupation, is by no means new. As far back as 1751, d'Alembert acknowledged that having found here and there a real advantage in the kind of knowledge where we had not sus-pected it, we felt authorized to consider all pursuance of mere curiosity as of possible future use. D'Alembert's testimony is important, for with him we are close to the positivistic scientist of the nineteenth century.[19]

We seem to have reached a peculiar impasse. Science, we reaffirmed (for the belief is very old), was the heir of the philosopher, willing to make sacrifices for the sake of the truth. We agreed that since the seventeenth century science increasingly acknowledged the element of usefulness, though this was in addition to, or side by side with, searching for the truth. Now we are forced to say that scientific curiosity, being a dedication to pleasure, insisted on being satisfied regardless of any useful results. To be sure, one can be dedicated to pleasure, but such dedication will hardly be called a virtue.

If serious philosophical curiosity can appear selfish, this could be even worse in the case of scientific curiosity. For the scientist separated him-self from the natural philosopher not only by renouncing religious and metaphysical issues. He fractioned even positive knowledge into circum-scribed problems, which he proceeded to solve in their isolation. Harvey's work on the circulation of the blood usually is cited as an early example of the new experimental science, which was satisfied with giving correct though partial solutions. The heart and its movements had been seen in connection with respiration. Harvey did not explain this nexus but left the answer to others. He is greatly praised for this, and rightly so, because his procedure made a solution possible. Later on, others supplied other pieces of the puzzle. This worked well, but there was hardly any assurance that it would always work well. At any rate, the pleasure of finding out more and more about less and less—as we would phrase it—could hardly be considered a particular merit, even if it proved useful. One might then say that science as the sum of individually pleasurable occupations repre-sented the pursuit of truth. If it did not work out, then curiosity would be the scientist's vice, and he would have to be ruled and regulated by pow-ers who saw to it that his curiosity took the right direction.

Having reached such unpleasant thoughts, I remember gratefully that my reflections are to be historical, and I look for a solution of the dilemma in another quotation, this time from John Locke. It was Locke's view that "he that would seriously set upon the search of truth ought in the first place to prepare his mind with a love of it." Here again we meet an emotion, but this time it is love rather than pleasure. And love for somebody or something else is, by definition, not quite as selfish as pleasure can be. Locke thinks that the test of whether a man really loves truth is his "not entertaining any proposition with greater assurance, than the proofs it is built upon will warrant."[20] This implies that the scientist must not only be critical but also be prepared to be proved wrong and that he must offer to other scientists the opportunity of so proving him. This is more than submitting to the rules of the game so as to prove oneself. Love brings pleasure, and the lover is willing to bring sacrifices for the beloved object.

With Locke's phrase, echoed by Polanyi,[21] Sarton, and many others, we have indeed found a way out of our impasse, though at a price. Love, it will be admitted, must be spontaneous; it is not subject to command. And what if the scientist does not find love of truth in his heart? I am here reminded of a story told about Benjamin Jowett, the translator of Plato and the powerful master of Balliol College, Oxford. Once an undergraduate student said to him that he could not see God in the universe and could not find him in his heart either. Whereupon Jowett commanded, "You must either find Him by tomorrow morning, or leave the College."[22] This was a harsh order, and the order would remain just as harsh if God were replaced by the love of truth.

Having envisaged a number of ways in which the scientist is related to truth, we shall pursue that which is marked by dedication or love—not because this relationship is necessarily the only possible one, but because it is the one that implies virtue. Curiosity and utility are not excluded. We assume that the scientist is curious to *know*, that is, that he wants to be able to say, "This result of mine is the true solution." We also assume that scientists investigate innumerable isolated problems with methods worked out in ever finer detail and reliability. But we add that the scientist dedicated to truth is not indifferent to what goes beyond his limited specialization, that it matters to him whether and how his own work will relate to that of others, now and in time to come, and what all scientific endeavors amount to. In other words, we assume that the scientist looks

upon himself as a member, and not only a particle, of science and that he hopes that in his work science as a whole is illuminated. This, I think, is a fair picture of the scientist of the early nineteenth century, as he saw himself and as he was represented by the great scientific associations.

In different countries the emancipation of science from philosophy and theological dogma proceeded differently. In France this emancipation had already taken place at the end of the eighteenth century. In England, on the other hand, a scientist like William Lawrence in the 1820s was silenced when he tried to discuss mental functions in purely physiological terms. Lawrence's defeat was not caused by his religious adversaries alone; it was helped by his own inability to decide between the claims of scientific *truth* and religious *faith.*[23]

In Germany the situation was again different. Natural science, *Naturwissenschaft,* was just one aspect of *Wissenschaft* in general, a term that embraced all methodical studies. In the early decades of the century, *Wissenschaft* was led by an idealistic metaphysics culminating in the system of Hegel. Around the middle of the century, idealistic metaphysics was replaced by a materialistic philosophy. Its most radical leaders, mainly scientists and physicians, claimed that matter and force were the ultimate reality and that all riddles of the world could be solved by science. In other words, scientific truth was the exclusive truth, and the pursuit of science was the pursuit of the whole truth. Science gave a degree of intellectual assurance that was just as great as the intellectual assurance that the Hegelian system had offered. It is against this background that two speeches, given by Rudolf Virchow and Emil Du Bois-Reymond, have to be understood.

On September 22, 1877, Rudolf Virchow spoke before the fiftieth conference of the German Association of Naturalists and Physicians on "the freedom of science in the modern state."[24] Noting with satisfaction that scientific teaching had become free, Virchow warned that this freedom must be safeguarded against internal abuses that might lead to its destruction. He drew a sharp line of demarcation between established scientific truth and mere hypotheses, such as the animal descent of man, the micro-organismic nature of all infectious diseases, and the occurrence of spontaneous generation in the beginning of organic life. The public, he argued, must not be taught as scientific doctrine what was not known to be "objectively true." How did Virchow conceive of the "objectively true"?

He spoke of "the domain which [science] has actually won and fully settled," of delivering to the nation "an established truth of science, which
is attested with certainty and concerning which there cannot remain the
least doubt," and he went so far as to demand of "a perfectly-established
doctrine" that it be "so certain that we could pledge our oath to it, so sure
that we could say 'Thus it is.'"[25]

Virchow's speech was clearly influenced by his involvement in the *Kul-
turkampf* (the term itself was Virchow's[26]), which the Prussian government
was waging against the Catholic Church, and the attempt to stem the rise
of the Social Democrats. He was afraid of exposing science to popular
favor or disfavor. For us, the essential point lies in the fact that Virchow,
consciously or unconsciously, had a mental picture of science as the owner
of a storehouse of proven knowledge forever safe from serious doubt and
forever increasing in size. This was "established scientific truth."

In this respect he differed from his German colleague Du Bois-Reymond, for whom science had its clear-cut goal. Scientific knowledge meant
the *reduction* of the changes in the corporeal world to movements of atoms,
such movements being caused by the atoms' central forces, which were
independent of time. Conceivably, some day the whole cosmic process
would be represented "by one mathematical formula, by one immense system of simultaneous differential equations, from which would follow position, direction of movement, and velocity of every atom in the universe
at any time."[27]

Du Bois-Reymond was a reductionist who firmly believed that all scientific knowledge, including biology, had to be based on analytical mechanics of the classical type. He had said so in 1848, and he said so again
in 1872 in his famous paper "On the Boundaries of Scientific Knowledge,"
from which the above quotations were taken. Both Du Bois-Reymond and
Virchow warned against an undue extension of science. But whereas for
Virchow the danger lay in blunting the boundaries between assured scientific truth and hypothetical scientific beliefs, Du Bois-Reymond went
on to stress the boundaries of science as such. Though natural processes
were reducible to atoms and their central forces, the essential nature of
matter and force was incomprehensible and their existence had to be presupposed. Though life could be explained mechanically, consciousness
could not. These limitations of science led Du Bois-Reymond to close
with the word *Ignorabimus* [we shall remain ignorant].

The strong impression that Du Bois-Reymond's declaration made on his contemporaries is somewhat surprising in view of the fact that similar sentiments had been expressed in England years before by John Tyndall.[28] And of Thomas Huxley the *Spectator* had said in 1870 that in theory he was "a great and even severe Agnostic, who goes about exhorting all men to know how little they know."[29] But that was England, where metaphysics leaned less toward science than in Germany.

Du Bois-Reymond was not an empirical skeptic like Huxley, nor a French positivist for whom metaphysics belonged to a past stage of history. He admitted that the essential nature of matter and force, and the relationship of body and mind, posed legitimate problems. At the same time he removed them from the jurisdiction of science, as Pasteur removed religious faith from its jurisdiction. Nevertheless, Du Bois-Reymond, too, represented the universalistic claim of science to knowledge. As one critic had it, Du Bois-Reymond did not see how science could solve certain problems; hence he proclaimed them forever incapable of solution.[30]

Compared with both Virchow and Du Bois-Reymond, not to speak of the out-and-out materialists, Claude Bernard was less dogmatic and his attitude to truth much more complex. In spite of leanings toward positivism, Bernard recognized a metaphysical element in human thought that he wished to separate from strict science, rather than to suppress or throw into the limbo of unsolvable riddles.[31] Scientific truth was relative; in this limitation lay its strength. In all experimental science, "we can reach only relative or partial truths and know phenomena only in their necessary conditions. But this knowledge is enough to broaden our power over nature."[32] Claude Bernard was a physician, anxious to create a scientific basis for curative medicine. Truth might be relative, but this was balanced by the promise it held to benefit mankind.

In contrast to Virchow, Bernard would probably have refused to take his oath on any scientific truth. And in contrast to Du Bois-Reymond, he would probably have denied that analytical mechanics was forever unshakable. "Even if no observation has so far disproved the truth in question, still the mind does not therefore imagine that things cannot happen otherwise; so that it is only by hypothesis that we admit the principle as absolute."[33]

By relative truth Bernard meant that any hypothesis is true only insofar as it is supported by present experimental evidence. For most practicing scientists, and historians too, this probably is all they need in their

work, the ability to say: I can prove that this is so, or that this carries the greater probability. If tomorrow's proof shows otherwise, that is a matter for tomorrow. But when we look up from our daily work, we have to admit that relative truth still claims to be truth, and when we are in a philosophical mood we can hardly help asking how the many relative truths are connected. Are we to think of a process from hypothesis to experiment to hypothesis to experiment ad infinitum? If this were all, science would make little sense. It would obtain meaning only by our permitting ourselves to interrupt the chain at any point where practical applications can be obtained. Such a pragmatic view of science orients it toward human action subject to ethical criteria. These, rather than truth, become the final judge. By a chronological coincidence, the year of Claude Bernard's death, 1878, also witnessed the publication of Charles Peirce's essay marking the rise of pragmatic philosophy in America.

But whereas the scientific process can be thought of as extending into an unlimited future, it has no unlimited past. Its beginning rests on an orientation that was the product of definite historical periods. Historical universalism claims that different periods with different assumptions will lead to different modes of science. But this type of historicism, the twin brother of scientism, gained strength only in the twentieth century.

As far as Bernard is concerned, a third metaphysical attitude was probably closer to him. The process of hypothesis and experiment was pointing toward a distant truth that could no longer be challenged. As with the asymptotes of a hyperbola, the point of contact will never be reached, yet there is direction in the process as well as usefulness. It is hard to prove that Bernard envisaged science as progressing toward an unreachable absolute. Yet that is the impression given by his friend Ernest Renan, who, on succeeding to Bernard's seat in the French Academy, said:

> Truth was his religion; he never had any disillusionment or weakness, for not for a moment did he doubt science. Now science gives happiness whenever one is content with her and only asks of her what she can give. If she does not respond to all the questions which the greedy and the eager address to her, at least what she does impart is certain. The results of modern science are not less valuable for being acquired by successive oscillations. These delicate approximations, this successive refining, which leads us to modes of seeing *ever closer to truth* are the very condition of the human mind. Science thus gave to our colleague all the calm which the certitude

of being right procures. He envied nobody, he believed that his was the better part.[34]

I think that many people, scientists and laymen alike, have in their minds this picture of a journey toward distant, absolute truth, and I feel certain that George Sarton did. It was one element of his enthusiasm for the history of science. If only we would permit it, the march of science toward truth could also lead humanity toward a morally better life. Indeed, Sarton told us that "veracity, complete and unrestricted, is the conquest of science, even of modern science, and earlier people could have no conception of it."[35]

Sarton saw the discrepancy between what science should do and what it had accomplished. It had changed the surface of the earth, and, having provided a new standard of objectivity, it ought to lead man to veracity and social progress. Indeed, the progress of veracity "ought to be our measuring rod for the real scientific advance."[36] Unfortunately, this progress was as slow and precarious as was social progress. Greed and hypocrisy, Sarton thought, were responsible for the delay.

At this point I must confess to a feeling of embarrassment, which has accompanied me all through my talk. I have been speaking about truth with a capital T, so to say. This is as bad as talking about Nature or Soul. It is not the thing to do. The generation of Bernard, Huxley, and Pasteur was more robust in this respect than we are. They still had to fight for the right to scientific truth. We have inherited this right, and it is not seemly for the heir to strike the posture of the man who earned the fortune.

But I believe that we have a weightier reason for embarrassment than the fear of lacking due modesty. In the time of Bernard, Huxley, and Pasteur, just mentioned, psychology and the social sciences were still in their infancy. They have now come of age, and they have taught us a lesson about veracity. We have learned to be on our guard against spelling such words as truth with capital letters. Before accusing the world of not measuring up to scientific standards of truthfulness, we ought to consider whether the whole talk about dedication, sacrifice, love of truth is not really cant. I have already said that the scientist doing useful work insists on his pay. I admitted that scientific curiosity can be selfish. Now a skeptical voice asks whether there is any truth in saying that scientists work for the sake of finding the truth because they love it. Scientists are human be-

ings. They do the jobs they are best suited for, and they do them as best they can. They long for rewards and compete with one another, and some of them do administrative work, because they like power or because they cannot resist the pressure exerted on them. Beyond this, they are more likely to turn toward the social obligations of science than to love of truth for its own sake or for the sake of its beauty.

To be sure, this is an oversimplified and exaggerated description; nobody will claim that the motivation of all scientists can be put into such a nutshell. Nevertheless, the description helps to make clear the warning that the scientist's so-called virtue is an idealistic artifact with little relation to reality. If this were admitted, runs the argument, the world and the scientist would have made a step forward in veracity.

It would be foolish to deny that scientists conform to the description I just sketched: how many do and to what degree, I am unable to guess. But I do not believe that it holds true of all scientists at all times. If such were the case, science probably would not exist. At any rate, my theme is the scientist's virtue and not how many scientists live up to it.

Nevertheless, I have not tried to build up a straw man in order to knock him down, more or less successfully. In the first place, I am fully aware of the desirability of studying the scientist's way of life, and I, too, believe in the importance of some relationship of the scientist to his ideal, lest the latter lose all binding force. Only I think it is a matter of the stand taken by scientists as individuals. A last flight into the past will show what I mean.

In 1633, before the tribunal of the Holy Office in Rome, Galileo Galilei abjured the heliocentric system that he had supported in a recent book. Not being a Galileo scholar, I cannot comment on the event, which has been thoroughly treated by Professor Santillana.[37] But I have been intrigued by the well-known legend according to which Galileo, having recanted, said "eppur si muove" [and yet the earth moves]. The legend expresses what, in the judgment of posterity, truth would have forced the scientist to mumble, or at least to think. In 1796 Laplace, in his popular *Exposition du système du monde*, wrote, "What a spectacle!—an old man, famous for his long life dedicated entirely to the study of nature, abjuring on his knees, against the testimony of his conscience, the truth he had proved on the basis of evidence!"[38]

Hegel cited this passage in illustration of the difference between the church and the state. The church as the guardian of subjective truth—that is, faith—could turn against science, which the state was bound to defend. The purpose of science was rationality and knowledge—that is, "the thinking of objective truth"—and this coincided with the interest of the state, whose authority must be able to defend itself against subjective claims.[39]

The agreement between Laplace, the exact scientist and ideologue, writing in revolutionary France of 1796, and the metaphysician Hegel, lecturing in Berlin, the center of conservative, monarchic Prussia, is significant. It points to some degree of standardization in judging Galileo among nonclerically oriented Europeans of the nineteenth century. Though Prussia suppressed all democratic tendencies, in purely scientific matters it could be more tolerant than England. But we have since learned that Hegel erred and that intolerant states and societies can be as suppressive as any Inquisition.

At any rate, the image of Galileo as the martyr of science has survived. A crime was committed against a great scientist; he recanted because he had to yield to force.

While authorities of the church have now expressed regret over the treatment meted out to Galileo, Santillana as well as Berthold Brecht has deemphasized the specific role played by the Inquisition,[40] which has become the historical representative for any suppressive authority, and Santillana has drawn parallels with modern events. Santillana has also drawn our attention to the bewilderment in which Galileo found himself as a faithful Catholic.

Without any systematic search, I have encountered differing opinions expressed by several well-known men, which I shall briefly list without any regard to whether they are right or not or to mutual dependencies or chronological relations. According to E. J. Dijksterhuis, Galileo, fortunately for all concerned, did not behave like a martyr and should not be seen as one.[41] In C. F. von Weizsäcker's opinion, Galileo had no real proof for the heliocentric system, and his fanaticism in supporting it had justification only insofar as science needs daring assertions if it is to advance.[42] Galileo's cosmological work, which caused the trouble, struck Arthur Koestler as a distorting piece of propaganda; Galileo's trial had the ef-

fect of exposing him and bringing him back to work on matters he understood.[43] This was in direct contrast to Brecht's interpretation of Galileo's renunciation as a crime against the kind of science that can set people free from their oppressors.[44] Karl Jaspers thought that a demonstrable truth, like the heliocentric system, does not need sacrifice, which is demanded only where conviction is vested in personal faith.[45] Finally, Albert Camus, too, commented on Galileo's ready abjuration of an important scientific truth as soon as his life was endangered. In a certain sense, Camus thought, Galileo was right. "This truth was not worth the stake. Which of the two, earth or sun, rotates around the other, is profoundly indifferent. To say it straight out: it is a futile question."[46]

Quot homines tot sententiae, "as many opinions as there are men," one might say! Yet one thought stands out: Galileo is not just a scientist versus the Inquisition; it is this particular man and his crime or innocence and, on the other hand, it is the relation of scientific truth to human values that preoccupy us and prevent the case of Galileo from coming to rest. Unless I am mistaken, this suggests that we suffer from some unresolved doubts. The nature of the doubt reveals itself if we take the above opinions simply as possible ways of behaving in a similar situation. If any one of us were to find himself in Galileo's situation, how would he behave?[47] Would he excuse himself by pleading force majeure? Would he say *eppur si muove?* Would he become convinced that the power accusing him was right and that he was wrong? Would he discover that his scientific findings really were not certain enough to allow him to take a stand? Would he believe that the truth of his results was so strong that it did not need his sacrifice? Or would it appear senseless to him to let any relative scientific truth affect his personal fate? We can only hope to be spared such an ordeal. If we were to experience it, each one would have to make his own decision; I do not know of any rules or guidelines that could be relied upon, and this, indeed, is embarrassing.

But we need not end on such a macabre note. Without seeing ourselves in a dungeon,[48] we can take the case of Galileo as a means of realizing that the scientist's virtue, if it is to be a moral force, cannot be entirely detached from the person who is the scientist. For better and for worse, among other things science has become such a great power in our world that its organization, support, and wise direction are matters of national and international necessity. This is a truism. What should not be forgotten is

another truism: that organization, support, and even wise direction do not substitute for the inner life of science, that is, the scientists who are human beings and not only a quantity of talent to be trained, supported, and surrounded by social prestige. For this reason I have dwelt on dedication, sacrifice, pleasure, love, veracity, doubt, even on man as an *animal metaphysicum*, as Schopenhauer had it,[49] for all these are human factors. If doing so needs an excuse, then the name of George Sarton, who strove so valiantly for the humanization of science, will be excuse enough.

NOTES

This chapter began as the 1968 George Sarton Memorial Lecture, delivered December 28, 1968, in Dallas, Texas, during the annual meetings of the History of Science Society and the American Association for the Advancement of Science.

1. George Sarton, *The Life of Science: Essays in the History of Civilization* (New York: Schuman, 1948), 40, 55.

2. Ibid., 164.

3. E.g., see Sarton's *A History of Science: Ancient Science through the Golden Age of Greece* (Cambridge: Harvard Univ. Press, 1952), chap. 16, exp. 408 ff.

4. Plato, *Phaedrus* 270 C–D.

5. Plato, *Republic* 10. 595 C. See also *Phaedo* 91 C, where truth is put above Socrates.

6. Aristotle, *Nicomachean Ethics* 1.6.1.1096a.14 ff. (H. Rackham's trans., Loeb Classical Library, p. 17).

7. For the provenance of this adage, see Klaus Bartels and Ludwig Huber, *Veni, vidi, vici. Geflügelte Worte aus dem Griechischen und Lateinischen*, 2nd ed. (Zurich: Artemis, 1967), 70.

8. Galen, *Hoti ho aristos iatros kai philosophos*, ed. Iwan Mueller, *Galeni Scripta minora*, vol. 2 (Leipzig: Teubner, 1891), chap. 3, pp. 6, 9–10; chap. 4, pp. 8, 10–11. I omit those virtues which, according to Galen, the physician must have *qua* healer.

9. Galen, *On Anatomical Procedures*, bk. 1, chap. 3, ed. C. G. Kühn, 2:231–34; trans. Charles Singer (New York: Oxford Univ. Press, 1956), 7–8.

10. Annie Barnes, *Jean Le Clerc (1657–1736) et la république des lettres* (Paris: Droz, 1938), 13: "Mais la conception d'un République des Lettres ne semble pas remonter plus haut que la première moitié du XVIIe siècle, et l'idée n'en devient populaire qu'autour de 1700." See also H. R. Trevor-Roper, *Religion, the Reformation, and Social Change* (London: Macmillan, 1967), 202.

11. *The Works of William Harvey, M.D.*, trans. Robert Willis (London: Sydenham Society, 1847), 7, 149; cf. *Opera omnia* (London, 1763), 165.

12. See the title pages of Galileo's various works. On the Accademia dei Lin-

cei, see Martha Ornstein, *The Role of the Scientific Societies in the Seventeenth Century* (New York, 1913), 91–92.

13. Aristotle, *Metaphysics* 1.1.981b.

14. Pardon E. Tillinghast, *Approaches to History* (Englewood Cliffs, N.J.: Prentice Hall, 1963), 92–95.

15. Thomas Hobbes, *Leviathan*, pt. 1, chap. 6, in *English Works*, ed. Molesworth (London: Bohn, 1839–45), 3: 44 f.

16. Lewis S. Feuer, *The Scientific Intellectual: The Psychological and Sociological Origins of Modern Science* (New York: Basic Books, 1963).

17. William James, *The Principles of Psychology*, 2 vols. (reprint, New York: Dover Publications, 1950), 2: 429 f.

18. Arthur O. Lovejoy, "Reflections on the history of ideas," *J Hist Ideas* 1 (1940): 3–23, see 19.

19. Jean d'Alembert, "Discours préliminaire des éditeurs," in *Encylopédie, ou Dictionnaire raisonné des sciences, des arts et des métiers, par une société des gens de lettres*, new ed. (Geneva: Pellet, 1777), 1:x; see also Frank E. Manuel, *The New World of Henri Saint-Simon* (reprint, Notre Dame: Univ. Notre Dame Press, 1963), 145. Shortly before d'Alembert, J. J. Rousseau in his *Discours sur les sciences et les arts*, ed. George Havens (New York: Modern Language Association of America, 1946), 129, had said: "L'Astronomie est née de la superstition; l'Eloquence, de l'ambition, de la haine, de la flatterie, du mensonge; la Géométrie, de l'avarice; la Physique, d'une vaine curiosité; toutes, et la Morale même, de l'orgueil humain."

20. John Locke, *An Essay Concerning Human Understanding*, ed. John W. Yolton (Everyman's Library), vol. 2, bk. 4, chap. 19, p. 288.

21. Michael Polanyi, *Science, Faith, and Society* (London: Geoffrey Cumberlege, Oxford Univ. Press, 1946). Polanyi stresses the implications contained in the love of science.

22. Gilbert Highet, *The Art of Teaching* (reprint, New York: Vintage Books, n.d.), 203.

23. Owsei Temkin, "Basic science, medicine, and the Romantic era," *Bull Hist Med* 37 (1963): 97–129.

24. Rudolf Virchow, *The Freedom of Science in the Modern State*, 2nd ed. (London: Murray, 1878). A reply was published by Ernst Haeckel, *Freie Wissenschaft und freie Lehre* (Leipzig: Kröner, 1908).

25. Virchow, *Freedom of Science*, 63, 8, 19.

26. Erwin H. Ackerknecht, *Rudolf Virchow: Doctor, Statesman, Anthropologist* (Madison: Univ. Wisconsin Press, 1953), 185.

27. Emil Du Bois-Reymond, "Ueber die Grenzen des Naturerkennens," in *Reden*, 1st series (Leipzig: Veit, 1886), 105–40; see 107.

28. John Tyndall, "Professor Virchow and evolution," in *Fragments of Science* (New York: Appleton, 1892), 2:373–418; see 392.

29. Quoted from *A New English Dictionary*, s.v. "agnostic."

30. Heinrich Rickert, *Die Grenzen der naturwissenschaftlichen Begriffsbildung*, 2nd ed. (Tübingen: Mohr, 1913), 7.

31. On Claude Bernard's attitude to metaphysics, see D. G. Charlton, *Positivist Thought in France during the Second Empire, 1852–1870* (Oxford: Clarendon, 1959), chap. 5, and Reino Virtanen, *Claude Bernard and His Place in the History of Ideas* (Lincoln: Univ. Nebraska Press, 1960), 17, 49 ff., 114, 132.

32. Claude Bernard, *An Introduction to the Study of Experimental Medicine*, trans. H. C. Greene (New York: Schuman, 1949), 82.

33. Ibid., 30. Bernard is speaking of the principles of theoretical mechanics and "some branches of mathematical physics" (29).

34. Ernest Renan, "Claude Bernard," in Renan et al., *L'Œuvre de Claude Bernard* (Paris: Baillière, 1881), 3–37; 33 (italics are mine).

35. Sarton, *The Life of Science*, 161, 183.

36. Ibid., 183.

37. Giorgio di Santillana, *The Crime of Galileo* (Chicago: Univ. Chicago Press, 1955).

38. *Œuvres complètes de Laplace* (Paris: Gauthier-Villars, 1884), 6:435.

39. G. W. F. Hegel, *Grundlinien der Philosophie des Rechts*, 3.3.269, in *Sämtliche Werke*, ed. H. Glockner (Stuttgart: Frommann, 1928), 7:359 f.

40. Santillana, *Crime of Galileo*; Bertolt Brecht, *Leben des Galilei*, pt. 8 (Frankfurt: Suhrkamp, 1962), 206 f.

41. E. J. Dijksterhuis, *The Mechanization of the World Picture*, trans. C. Dikshoorn (Oxford: Clarendon, 1961), 62.

42. C. F. von Weizsäcker, *The Relevance of Science* (New York: Harper and Row, 1964), 110 f.

43. Arthur Koestler, *The Sleepwalkers* (New York: Macmillan, 1959), 476 f., 494 f.

44. Brecht, *Leben des Galilei*, 204.

45. Karl Jaspers, *Der philosophische Glaube* (reprint, Frankfurt: Fischer Bücherei, 1958), 11.

46. Albert Camus, *Le mythe de Sisyphe: Essai sur l'absurde* (reprint, Paris: Gallimard, 1967), 15 f.

47. The Lysenko affair has, of course, presented such a test case; see Bentley Glass, *Science and Ethical Values* (Chapel Hill: Univ. North Carolina Press, 1965), esp. 82–92. I have preferred the example of Galileo because its historical distance has provided a spectrum of opinions relatively free from modern ideological ties. For a defense of Galileo against Koestler, see Feuer, *The Scientific Intellectual*, 163.

48. Which Galileo was also spared.

49. Arthur Schopenhauer, "Ueber das metaphysische Bedürfniss des Menschen," chap. 17 in *Die Welt als Wille und Vorstellung* (Leipzig: Hesse and Becker, 1919), 2:195.

The History of Therapy
and Nutrition

CHAPTER 8

Historical Aspects of Drug Therapy

It would be a fruitless enterprise to enter upon speculations on the origin of drug therapy.[1] Suffice it to say that it is very old. A Sumerian clay tablet recording a number of recipes (though not telling for what) has been dated as far back as 2100 B.C.[2] Whether this document is altogether free from magic or whether accompanying magic formulas were supplied from other texts is a moot question, for there is no lack of Sumerian exorcisms to cure sick people.[3] Ancient Mesopotamian and Egyptian texts lead us to believe that, at some time at least, the use of drugs was imbued with magic thought. In the Homeric epics the term *pharmakon* connotes a charm or a drug that can be used for good and bad purposes.

To say that a drug has magic associations means that its action for good or bad does not depend on its natural qualities alone. The favor of a god, the observance of ceremonies, the absence of demoniac enemies, the healing intent of the dispenser—any or all of these factors are needed to make it therapeutically effective.[4]

A different attitude prevailed in the writings of the Greek physicians from the Hippocratic times of about 400 B.C. until the end of antiquity.[5] Culminating in the work of Galen, the attempt was made to establish a physiological explanation of drug action. In the final analysis, the contrary qualities of dry and wet, of cold and hot, were believed to determine the condition of the human body in health and disease. Too much or too little of these qualities in the solid parts of the body or in its fluids would produce the various diseases. To recognize the nature of the disease in the

Originally published in *Drugs in Our Society*, edited by Paul Talalay (Baltimore: Johns Hopkins University Press, 1964).

individual patient and to counteract it by contrary measures was the task of the physician. Individual factors like constitution, age, or season made it necessary to individualize the diagnosis, so that every illness had something unique. Therapeutic measures likewise had to be individualized. As to drugs, their qualitative composition could be investigated, and Galen formulated principles of experimental procedure,[6] which were developed in the Middle Ages into a logic of experimental science.[7] The Galenic physician in the Middle Ages and Renaissance believed himself in possession of a science of pharmacodynamics that was to help him in selecting the right drugs in right proportions for the treatment of his individual patient. The drugs were tools that the learned physician alone could use, if and when necessary. This distinguished him from the apothecary, who knew the drugs but not humans, from the empiric, who at best had learned by experience of some remedies efficacious against certain symptoms or disease entities.

In some respects the rationality of this system was more pretended than real. The turn from magic to naturalism was neither as abrupt nor as complete as it might appear. The word *pharmakon* in the Hippocratic writings, apart from meaning remedy or poison, also has the specific connotation of a cathartic drug. The cathartic, or literally "cleansing" drug, was used to cleanse the body of superfluous humors and peccant matter.[4] But this idea of cleansing was related to the older magic idea of cleansing a person of impurities that defile him in the sight of the gods and of man.[8] What on the surface looks like a rational etiological procedure turns out to be underpinned by magic elements. This may go far in explaining the persistence of the use of cathartics down into our own century.

Much as the Galenic system stressed the physiological basis of drug action, it admitted a class of drugs that were effective in certain cases, although nobody could say why. This was a concession to purely empirical therapy, and it opened the doors wide for all kinds of superstitious remedies, including animal organs and excrements. Once the appeal to mere experience was admitted, a safeguard was needed against the empiric (i.e., the quack). This safeguard was found in tradition and in the reputation of the physician. The possession of an academic degree and the reliance on remedies that had been used for centuries were the credentials of the doctor as he emerged in the Middle Ages.

When the fight began in the sixteenth century between the Galenists

and the followers of Paracelsus—who insisted on the internal use of chemically prepared drugs—more than the use of these drugs was involved. As Multhauf showed, chemical drugs had slipped into the materia medica before Paracelsus.[9] But with him there arose the claim that experiment was more valid than tradition, that therapy must utilize the arcanum, the potent quality of drugs to be obtained through chemical processes. Paracelsus's contempt for the traditional pharmacopoeia and his insistence on new modes of preparing drugs, often from substances considered poisonous, postulated much more than a scientific and philosophical reform. Paracelsus was not a reformer; he was a revolutionary threatening the social existence of the medical profession.

By the beginning of the eighteenth century, the battle between Galenists and Paracelsists was about ended with the disappearance of the former and the reduction of the latter to a small minority. Galenism as a scientific system had succumbed to the scientific revolution of the seventeenth century. Chemical drugs were here to stay, and the apothecary who prepared them in his laboratory emerged as the leading chemist of the eighteenth century. The materia medica was being enlarged by chemicals such as calomel and Glauber's salt, by imported drugs such as cinchona and ipecacuanha, and by native herbs like foxglove. During the Enlightenment it was also purged of things that smacked of magic.

As a scientific system, Galenism had been overthrown, but it had not been replaced by generally acceptable therapeutic principles, especially as far as treatment by drugs was concerned. If one wants to characterize the situation around 1850, *chaotic* is probably the most appropriate term.[10] Side by side there existed general skepticism, sectarianism, and empiricism of varying shades and degrees.

Empiricism, to a larger or smaller extent, relied on the traditional materia medica. In Cullen's great work, which appeared in the late eighteenth century and remained paradigmatic far into the next century, the substances were arranged "according to their agreeing in some general virtues" whereby they were made to answer general therapeutic indications.[11] Hence we find the astringentia, tonica, stimulantia, sedantia, and other categories, which told the practitioner what to choose from if his patient needed to be stimulated, strengthened, sedated, and so on. This type of classification, which had its origins long before Cullen, is too well known to need elaboration.

Every practitioner might, of course, add drugs or prescriptions that he claimed to have found efficient. Fashion led to the rise and fall of drugs. For instance, far into the middle of the nineteenth century, mercury in general and calomel in particular were most popular among Anglo-Saxon physicians in the cure of inflammatory conditions. Second only to depletion of the blood, or instead of it, thought Dr. George B. Wood of Philadelphia, "no remedy has so powerful an antiphlogistic influence as mercury, urged to the point of affecting the system" (i.e., mercurial poisoning). The action of the drug was believed to be alterative, which meant that it changed "existing morbid actions or states," the mode of action being unknown.[12] The use of this kind of drug therapy by regular physicians contributed greatly to the popularity of the sects: the homeopaths, Thomsonians, eclectics, and osteopaths.[13]

> Your Calomel, and all your deadly drugs, reject!
> The world is wakening round you! Botanic
> Doctors (sounding the majesty of truth)
> Gain ground: the mercurial craft declines!
> Thick darkness flies before Thomsonian light,
> Bursting in glory on a long benighted world![14]

Neither Cullen nor his successors were naive believers in drugs. Cullen knew that his own system, like all therapeutic systems, was bound to physiological and pathological notions.[15] Moreover, he did not fail to discuss the rules whereby the virtues of medicines could be ascertained. Even experimentation on animals by feeding of drugs or injecting them into the veins was duly mentioned, together with its limitations in view of the differences between animals and humans. Wood, in principle at least, admonished physicians and laymen alike against the fallacy of *post hoc ergo propter hoc*,[16] though in practice he proceeded differently.

The whole relationship between belief and disbelief in drug therapy is difficult to disentangle because of the interaction of personal temperament, social factors, and scientific considerations. In antiquity Herophilus is said to have called drugs "the hands of gods"[17] and not to have treated any disease without them.[18] Asclepiades, on the other hand, believing that most drugs harmed the stomach, preferred dietetic treatment.[19] In the eighteenth century the great Leyden doctor, Gaubius, expressed doubt about all medical therapy.[20] Such general doubts were not

too frequently met on the part of the medical profession. On the other hand, laymen through the centuries scorned or ridiculed the doctors and their work. But there was a difference in the attitude of laymen and medical men. The laymen distrusted medicine, though not necessarily drugs, especially botanicals. Doctors defended medicine in one form or another but from the late eighteenth century on developed increasing skepticism toward the therapeutic efficacy of drugs. Apart from the difficulty of telling how far this became an individual matter and how far a matter of tradition, there is the added difficulty of deciding how far it was echoed by men in private practice. There is little doubt that skepticism was fostered in Vienna's *Allgemeines Krankenhaus* and in the large hospitals of Paris.[21] It is less certain that private practitioners without hospital appointments shared it. In the United States, where private practitioners set the tone in professional matters, the sentiments of Paris were echoed but hardly by many.[22]

At any rate, the skepticism as it grew up in Paris and Vienna and spread elsewhere was directed against bleeding (which we shall neglect here) and the traditional materia medica. Alleged experience was distrusted and was to be replaced by reliable observation, especially Louis's numerical method. The action of a drug was to be investigated relative to mortality and the slow or rapid progress of the disease. This presupposed a comparison of a rather large number of patients showing the same disease to the same degree, with some left untreated while others were administered this or that medicament. The same therapeutic agent had to be studied again in patients in whom the disease was severe and in those in whom the disease was moderate, in strong or weak doses, at a period close to or distant from the beginning, alone or together with other remedies.[23] The results were not very encouraging in Paris or in the hands of Skoda, who followed the same method in Vienna.[24] Experiments of this kind frequently showed an equal effect of quite different drugs given with the assurance of a cure.[25] The inference to be drawn was that any alleged effect of the drug lay in its placebo effect (to use a word now fashionable), an idea that had occurred long before regarding the effects of amulets and other superstitious remedies.[26]

Influenced by Louis's ideas, Bigelow in 1835 read a paper "On Self-Limited Diseases,"[27] in which he argued that a number of diseases were not influenced in their course toward recovery or death by the remedies applied.

At the same time Bigelow turned against a defeatist attitude in therapeutics. Even if the disease could not be cured, harm could be prevented and symptoms alleviated.[28] This was a prevailing attitude among physicians who believed that the recovery from disease was due to the healing power of nature, whose servant the physician ought to be.[29]

Medical men who wished to build pathology and clinical science on physiology had an additional reason for skepticism. The drugs might or might not be helpful for particular symptoms. But was it reasonable to expect them to act against diseases? Diseases were physiological processes existing in sick individuals only. Diseases were not metaphysical entities that lurked behind their manifestations and had to be subdued by drugs acting specifically against them. In the seventeenth century, ontology of this kind had been proclaimed by Thomas Sydenham. Around 1840, medical ontology once more was represented by the German school of "natural history," which explains why rising German physiological medicine was particularly outspoken in its fight against such notions.[30] "On one side stands the hostile demon of disease menacing the organism with its parasitism, on the other side the helpful healing power of nature! Thus do the mystics of medicine envisage the organism as an arena for their good and evil principles."[31]

Of course, Wunderlich, the author of this passage, knew that there was something "specific" in diseases like typhoid, cholera, syphilis. But it was not the disease entity, it was "the same specific cause which effects the same kind of reaction of the organism and the same kind of form of illness."[32] Lest there be any misunderstanding, Wunderlich was not looking for a micro-organism as the specific cause. The bacteriological interpretation of specificity did not emerge victoriously until forty years later.

Rational therapy meant therapy based on the insight into pathological processes; there was far-reaching agreement on this point.[33] There also was far-reaching agreement that a truly scientific etiology was still a postulate, a program very far from realization. In the meantime, therefore, two courses were left: to do nothing or to compromise with empiricism. I have the impression that so-called therapeutic nihilism was rejected by most physicians who had to manage patients. The remaining alternative was to compromise with empiricism. The compromise could vary in degree from polypharmacy to the avoidance of drugs as far as this seemed possible. To quote Wunderlich once more, who, in spite of all his skepticism, was hos-

tile to the therapeutic nihilists: "There is no kind of disease that cannot be cured without so-called medicaments and where these cannot be completely replaced by the thousand other remedies which are at the disposal of the rational physician." Which exactly the "thousand" other remedies were does not become clear. In practice, it had to be admitted that drugs, though often more than useless, had to be prescribed because of the superstition of the patient and to strengthen his confidence."[34]

Indeed, the people insisted on the use of drugs, even if they distrusted physicians and the particular drugs used by many of them. To the experienced practitioner this was quite clear. William Lawrence, one of those who as early as 1818 wished to build clinical medicine on the basic sciences, admitted that "a firm faith in drugs and plasters, and a liberal administration of them, may be the surer road to popular success, if the remark addressed by a veteran practitioner to a young enthusiast in science be well-grounded: 'Juvenis tua doctrina non promittit opes: plebs amat remedia.'"[35] [My boy, your teaching does not promise wealth: the crowd loves medicines.]

If there could be any doubt on this point, the sects, so abundant at the time, were likely to dispel it. True, the hydropaths and the osteopaths, in the youth of their existence, did without drugs. But the most widespread sects, around the middle of the nineteenth century, were the homeopaths and, in the United States, the various kinds of botanical doctors. As is well known, the homeopaths dispensed the drugs in such infinitesimal dilutions that the unbelievers considered them quite ineffective and homeopathic treatment equal to abandoning the patient to the healing power of nature. The various botanical doctors, though opposed to mineral drugs and particularly to calomel, nevertheless used drugs. Behind their predilection for plants, especially native ones, there was something of the old belief, voiced by Sydenham, that God had made provision "for the cure of the more serious diseases which afflict humanity, and that near at hand and in every country."[36]

In the preceding pages I have tried to picture the situation of drug therapy around the middle of the nineteenth century. This period was not only chaotic as far as the therapeutic use of drugs was concerned but also critical as far as the general relationship of healing to the science of medicine was concerned. Beginning around the 1830s, an increasingly large number of doctors of medicine severed their relation with the art of heal-

ing to devote themselves full time to subjects that did not bring them in contact with patients. To be sure, in previous times also some physicians had renounced medical practice to cultivate a particular science. But the regular separation of basic scientists from the practice of medicine as it began in Germany in the nineteenth century was a different matter. Carl Ludwig, Du Bois-Reymond, and Helmholtz were physiologists, not physicians. This is not the place to go into the details of this change in medicine, which had such far-reaching results. We need merely be concerned with the phenomenon that there existed men who had studied medicine, held medical degrees, usually were licensed to practice medicine, were members of the medical faculties of the universities, yet neither engaged in healing nor, in many cases, had any interest in it. Whether or not they justified their attitude by referring to a necessary division of labor is relatively unimportant. Within this broad movement, men, like Dietl, who are usually cited as extreme therapeutic nihilists represented clinicians led by similar sentiments or reasons to an abdication of active therapy at the bedside. They claimed that knowledge had to be gathered before healing could hold promise.[37] The quest for knowledge, regardless of its applications to practical problems, was the reforming spirit of the time.

This is the background for the transformation of the study of materia medica to a study of pharmacology. Scientific skepticism pointed to the ignorance regarding the physiology of disease and the physiological action of drugs. If the science of drugs was to emulate the other disciplines, it had to become a physiological science, too. In 1849, Buchheim closed an article called "On the Tasks of Materia Medica" with the words, "But many a 'ceterum censeo' may be needed to rouse pharmacology from its slumber. Incidentally, this is not a natural sleep, for pharmacology's achievements up to the present are no reason for being tired."[38] Buchheim was an important figure. He was the teacher of Schmiedeberg, who in turn brought John Jacob Abel to pharmacology. At first, Buchheim met with considerable resistance on the part of medical practitioners.[39] Yet in the long run his program was to gain recognition.

Pharmacology was to be a theoretical science that could be considered a part of physiology. On the one hand its task was to establish the active substances within the drugs, to find the chemical properties responsible for the action and prepare synthetically drugs that were more effective. On

the other hand, it had to study the changes brought about by the drug in the organism and then explore the possible influence of such changes upon pathological conditions.[40] It took its cue from the therapist, and it turned its results over to him.

In many respects this program as formulated in 1876 but summarized developments over a lengthy period. It presupposed the rise of modern chemistry, and it leaned heavily on the work of the great chemists, pharmacists, and physiologists, who had shown that the demands made were not utopian. In the course of the preceding sixty years, a number of potent substances had been isolated, such as morphine, codeine, emetine, iodine, bromine, strychnine, quinine, physostigmine, pilocarpine, and, above all, ether and chloroform. In 1821 Magendie, a pioneer of experimental pharmacology, had taught in his *Formulaire* the preparation and the use of some of the newly discovered active drugs.[41] Magendie's work had been continued brilliantly by his pupil Claude Bernard, whose investigation of curare had set an example for the physiological investigation of a drug. It is very important to realize the progress made during the time of prevailing skepticism and empiricism.[42]

This progress not only offered new drugs, but also had far-reaching consequences in another direction. These drugs could be manufactured in bulk and then simply sold in pharmacies. In France this led to a gradual decline of the status of the independent apothecary.[43] In Germany, on the contrary, it led to the rise of the chemical pharmaceutical industry. H. E. Merck, who started as an apothecary in 1816, in 1827 announced wholesale production of morphine, quinine, emetine, strychnine, and other drugs. Riedel acquired an apothecary's shop in Berlin in 1814; in 1827 he began to manufacture quinine. Schering started as an apothecary in 1851.[44]

But it would appear that this progress had not benefited the academic status of pharmacology. Buchheim's program was a reply to the great surgeon Billroth, who claimed that there was little work for the professor of pharmacology. All the latter could do was to give a survey of the most important groups of drugs, demonstrate the main types, and elucidate experimentally the effect of the most active poisons. The rest either belonged to the clinician or was a matter for pharmacists, medical examiners, and the like.[45]

While Buchheim was professor in Dorpat, he had to combine phar-

macology and the history of medicine, for neither discipline was thought to be full time.[46] When he died, materia medica had been dropped as an examination topic in German medical schools.

But Buchheim's program had its weaknesses. It was one thing to demand the correlation of the chemical constitution of the drug and its action, a correlation discussed by Brown and Fraser a few years before;[47] it was another thing to establish it successfully. Of more immediate concern, perhaps, was the range of drugs known and their therapeutic potentialities. What were the disease processes in which one could rationally and successfully intervene?

The germ theory had been proclaimed and was beginning to celebrate triumphs in antiseptic surgery. The whole subject of disinfectants thus became acute. But very few microbes had as yet been identified as causative agents of infectious diseases. Only with the advances of bacteriology and its cousin, immunology, would it become possible to devise a rational therapy of communicable diseases. Endocrine disorders likewise were not yet well understood. Anatomical diagnosis still prevailed over functional diagnosis to the disadvantage of disorders not approachable by surgery. In a state of medicine in which a major number of the disease conditions lacked proper understanding, it was perhaps too much to expect clarity in therapeutic indications. At any rate, as Ackerknecht showed in detail, even such drugs as digitalis, iodine, and quinine were far from being properly used.[48]

The analysis of the weaknesses of conditions around 1876, the year in which Koch opened the bacteriological era proper by his work on anthrax, implies events to come that would make pharmacology a powerful tool. It is impossible today to think of effective drug therapy without the help of the pharmaceutical industry. Now this industry, as we have seen, not only existed but, paradoxically, had already provoked complaints that sound relatively modern. Buchheim complained that:

> The chemical industry of our days produces various substances for which no market can as yet be found. Under these circumstances, the idea suggests itself that it might be possible to use these products as drugs. We know that a great number of physicians, without rhyme or reason, go after every new remedy that is recommended to them. If an industrialist is but shrewd enough to advertise sufficiently, he usually succeeds in increasing the sale of his product—for some time at least—and thus enriching himself.[49]

The complaint also reflects the relatively secondary place held by the manufacture of drugs in an industry mainly interested in dyes. The synthetic drug industry was just about to come into its own. Just then it was exploring the possibilities of antipyretics, especially the salicylates, and the century closed with the introduction of aspirin. But it also closed with the stage set for chemotherapy, after Ehrlich had explored the relationship between dyes and the living cells of the organism.

Having come close to the dawn of the modern era, I conclude with a disquieting thought, disquieting, that is, for the historian of medicine. It is becoming customary to look upon the history of drug therapy before chemotherapy and antibiotics as a prolongation of the Dark Ages. It is thought that by and large humanity would have been better off without it. Fortunate the patient who was not harmed by his physician. Most therapeutic successes were mere placebo effects. This view leaves but two heroes: the ancient believer in the magic action of drugs, with his trust in charms (i.e., the placebo effect), and the complete unbeliever, the therapeutic nihilist who prescribed drugs only as placebos, for the rest pinning his hopes on the progress of medical science. Is such a view justified?

I do not wish to appear here as a defender of the old materia medica, though I am not convinced that all was quite as bad and quite as vain as is often said. But even if we were to concede that the older history of drug therapy is of little interest regarding the therapeutic effect of drugs, we would still have to reckon with the history of those who made them, those who prescribed them, and those who took them. In short, there is still the history of the attitudes toward drug therapy. And this is the history of human beings perhaps not wiser but hardly more foolish than we are. And it is the history of times and circumstances that they shaped or to which they responded just as we do today.

NOTES

1. The following items should be consulted for this article as a whole: J. Petersen, *Hauptmomente in der geschichtlichen Entwicklung der medizinischen Therapie* (Copenhagen, 1877); H. Haas, *Spiegel der Arznei* (Berlin, 1956); E. H. Ackerknecht, "Aspects of the history of therapeutics," *Bull Hist Med* 36 (1962): 389–419. For interesting views on the beginnings of drug therapy, see W. Artelt, *Studien zur Geschichte der Begriffe "Heilmittel" und "Gift"* (Studien zur Geschichte der Medizin 23) (Leipzig, 1937).

2. M. Levey, *Chemistry and Chemical Technology in Ancient Mesopotamia* (New York, 1959), 149.

3. S. N. Kramer, *History Begins at Sumer* (Garden City, N.Y., 1959), 64.

4. For details see Artelt, *Studien.*

5. L. Edelstein, "Greek medicine in its relation to religion and magic," *Bull Hist Med* 5 (1937): 201.

6. O. Temkin, "Galenicals and Galenism in the history of medicine," in I. Galdston, ed., *The Impact of the Antibiotics on Medicine and Society* (New York, 1958): 18–37.

7. A. C. Crombie, *Robert Grosseteste and the Origins of Experimental Science, 1100–1700* (Oxford, 1953), 71 ff.

8. O. Temkin, "Beiträge zur archaischen Medizin," *Kyklos* 3 (1930): 90 (esp. 96 ff.). See also Artelt, *Studien,* 89 ff., and Ackerknecht, "Aspects" (esp. 391).

9. R. Multhauf, "Medical chemistry and the 'Paracelsians,'" *Bull Hist Med* 28 (1954): 101 (esp. 102).

10. A. Trousseau and H. Pidoux, *Traité de thérapeutique et de matière médicale* (Paris, 1855), cv. See also Temkin, "Galenicals," 30.

11. W. Cullen, *A Treatise of the Materia Medica* (Edinburgh, 1789), 1:158 ff.

12. G. B. Wood, *A Treatise on the Practice of Medicine,* 2nd ed. (Philadelphia, 1849), 1:214.

13. R. H. Shryock, *The Development of Modern Medicine* (New York, 1947), 252 ff.

14. Quoted from A. Berman, "The Thomsonian movement and its relation to American pharmacy and medicine," *Bull Hist Med* 25 (1951): 405.

15. Cullen, *A Treatise,* 153 f. For an account of the action of narcotics upon the muscles of animals, the heart "even of cold-blooded animals," and intravenous injection, see J. Murray, *A System of Materia Medica and Pharmacy,* with notes by N. Chapman (Philadelphia, 1815), 1:105.

16. Wood, *A Treatise,* 193.

17. Scribonius Largus, *Compositiones,* ed. G. Helmreich (Leipzig, 1887), 1.

18. Celsus, *On Medicine,* 5.1 (Loeb Classical Library), 2:2.

19. Ibid., 5.2.2.2.

20. I I. D. Gaubius, *The Institutions of Medicinal Pathology,* trans. C. Erskine (Edinburgh, 1778), vi. "It is not without concern, I must say, that I am in doubt whether mankind is more benefited or hurt by the medical art."

21. E. Lesky, "Von den Ursprüngen des therapeutischen Nihilismus," *Sudhoffs Arch* 44 (1960): 1; E. H. Ackerknecht, "Die Therapie der Pariser Kliniker zwischen 1795 und 1840," *Gesnerus* 15 (1958): 151.

22. Ackerknecht, "Aspects," 396.

23. P.-Ch.-A. Louis, *Recherches anatomiques, pathologiques et thérapeutiques sur la maladie connue sous les noms de gastroentérite* (Paris, 1829), esp. 2:459.

24. M. Neuburger, *Die Wiener medizinische Schule im Vormärz* (Vienna, 1921), 273 ff; B. Juhn, "Skoda und der 'therapeutische Nihilismus,'" *Ciba Symp* 3 (1955): 127; Lesky, "Von den Ursprüngen," and "Joseph Skoda," *Wien Klin. Wschr* 68 (1956): 726.

25. For instance, Esquirol's experiments on epileptics. O. Temkin, *The Falling Sickness* (Baltimore, 1945), 281.

26. Ibid., 102, 215, 227. In the eighteenth century such effects were ascribed to the power of "imagination."

27. In J. Bigelow, *Nature in Disease* (Boston, 1854), 1–58. Reference to Louis occurs on 34. Bigelow, 33, states that "hospitals and other public charities afford the most appropriate field for instituting them [i.e., observations] upon a large scale." W. Artelt, "Louis' amerikanische Schüler und die Krise der Therapie," *Sudhoffs Arch*, 42 (1958): 291.

28. Bigelow, *Nature*, 34 ff.

29. On this subject see Ackerknecht, "Aspects," 413 ff.

30. For German physiological medicine see K. Faber, *Nosography: The Evolution of Clinical Medicine in Modern Times*, 2nd ed. (New York, 1930), 59 ff.

31. C. A. Wunderlich, "Ueber die Mängel der heutigen deutschen Medicin," *Arch Physiol Heilkunde* 1 (1842): i.

32. Ibid., viii.

33. For a dissenting voice, see R. H. Lotze, *Allgemeine Pathologie und Therapie als mechanische Naturwissenschaften* (Leipzig, 1848), 45 ff. The origin of disease is not necessarily the target of therapy. A radical cure is nothing but the "complete removal of all inhibitions resisting the return of life to a permanent balance by means of the activity of the regulatory apparatus" (48).

34. C. A. Wunderlich, *Handbuch der Pathologie und Therapie* (Stuttgart, 1852), 75.

35. W. Lawrence, *Lectures on Physiology, Zoology, and the Natural History of Man* (London, 1819), 64.

36. *The Works of Thomas Sydenham, M.D.*, trans. R. G. Latham (London, 1848), 1:22.

37. On therapeutic nihilism see Petersen, *Hauptmomente*; Lesky, "Von den Ursprüngen"; Ackerknecht, "Die Therapie"; and H. Buess, "Zur Frage des therapeutischen Nihilismus im 19. Jahrhundert," *Schweiz Med Wochenschr* 87 (1957): 444.

38. Quoted from O. Schmiedeberg, "Rudolph Buchheim," *Arch Exp Pathol Pharmakol* 67 (1912): 6.

39. Ibid., 7.

40. R. Buchheim, "Ueber die Aufgaben und die Stellung der Pharmakologie an den deutschen Hochschulen," *Arch Exp Pathol Pharmakol* 5 (1876): 261.

41. J. M. D. Olmsted, *François Magendie* (New York, 1944), 79.

42. Ackerknecht, "Aspects," 396.

43. A. Berman, "The scientific tradition in French hospital pharmacy," *Am J Pharm* 18 (1961): 110, and "Conflict and anomaly in the scientific orientation of French pharmacy, 1800–1873," *Bull Hist Med* 37 (1963): 440.

44. A. Adlung and G. Urdang, *Grundriss der Geschichte der deutschen Pharmazie* (Berlin, 1935), 163 ff.

45. T. Billroth, *Ueber das Lehren und Lernen der medicinischen Wissenschaften an den Universitäten der deutschen Nation* (Vienna, 1876), 88–89.

46. Schmiedeberg, "Buchheim," 2–3. To be exact, Buchheim represented materia medica, dietetics, the history of medicine, and the encyclopedia of medicine.

47. A. C. Brown and T. R. Fraser, "On the connection between chemical constitution and physiological action," *Trans R Soc Edinburgh* 25 (1868–69): 151.

48. Ackerknecht, "Aspects," 400 ff.

49. Buchheim, "Ueber die Aufgaben," 271.

Galenicals and Galenism
in the History of Medicine

The history of the modern antibiotics is usually presented as following the rise of bacteriology and chemotherapy. This approach is undoubtedly correct insofar as modern antibiotics are defined as chemical substances obtained from an organism and antagonistic to a pathogenic organism.[1] The definition presupposes a knowledge of organisms as causative agents of disease, which means bacteriological knowledge. Besides, it presupposes a highly developed science of chemistry. The history of the antibiotics in the strict sense does not concern us in the present discussion.

Yet the antibiotic remedies have a wide historical background. Antibiosis is the antonym of symbiosis and means the antagonism of two organisms.[2] In this sense, any organism or a substance produced by it that harms another one acts as an antibiotic, regardless of whether we have any insight into the whole process. If somebody applies a colony of *Penicillium notatum* upon an infected wound, he uses antibiotic treatment, though he may be simply recommending the application of mold on a festering sore and be quite ignorant of the nature of the remedy and the process of infection.

In this sense, antibiotics have a long past. The interest roused by the miraculous effects of penicillin stimulated a search for older use of antibiotics and several instances have been recorded.[3] This example from Dioscorides' *Materia Medica* I came across the other day:

> But the rotten stuffe like meale, which is gathered out of old wood, and stocks of trees, being layd upon them, doth cleanse ulcers and bring them to cicatrix. It doth also stay Serpentia [spreading ulcers], being kneaded

Originally published in *The Impact of the Antibiotics on Medicine and Society*, edited by Iago Galdston (New York: International Universities Press for the New York Academy of Medicine, 1958).

together with the like quantitie of Anis and wine, as also being beaten small, put into linnen, and soe applyed.[4]

Dioscorides does not speak about the wood-rotting organisms involved, nor need we inquire here into the nature of the product or its possible efficiency if applied to infected wounds. It suffices to note that Dioscorides recommends rotting wood for the treatment of putrid sores and that Galen follows him with the words: "Rotten wood, and especially that which has an astringent and cleansing faculty like the elm tree, purifies and closes moist sores."[5] In the sixteenth century, Matthiolus, the famous commentator of Dioscorides, had this to add:

Indeed, the rotten part of old and wasting wood should not be entirely neglected since it possesses such a powerful faculty that it purges and heals up ulcers and stops serpentia. It becomes more efficient by being collected from the wood of trees that have an astringent as well as cleansing power. For this reason, rotten Guaiac wood easily excels in this respect, since it cures not only ulcers that yield easily, but also those which are supposed to be of an evil character and have been induced by the French disease. Thus if sprinkled on the corroding ulcers of the genitals, it heals them miraculously. Besides, not only the mould or rottenness of wood has come into medicinal use, but if we believe Pliny the worms too which originate in the rotten trunks of trees.[6]

Matthiolus probably was led to this antisyphilitic remedy because of the statements of Dioscorides and Galen and the widespread use of guaiacum in the sixteenth century. Since today we do not ascribe such powers to guaiacum, we may receive his praise with a good deal of skepticism. Whether or not the remedy in reality has any effect, we cannot tell without having tested it. But at the moment we are not interested in tracing possible antibiotics in the older literature. We merely wish to ask whether decaying wood, since recommended by Galen, may be called a galenical, and whether the same query relates to another possible antibiotic, yeast, since in a paragraph on leaven Galen says that "it draws up things from deep down and digests them."[7] Galenicals must obviously include the drugs used or recommended by Galen, the great physician of the second century, after whom they are named. On this basis, we have been dealing with two galenicals. If we use popular definitions, galenicals refer to vegetable simples or organic ingredients used alone or compounded with

other drugs in relatively simple preparations.[8] In that case we might even go so far as to include Fleming's penicillin broth of 1928. The chemical process of concentration and purification would then appear as the sole factor that takes antibiotics out of the class of galenicals. In principle, the difference would be about the same as that between cinchona bark and quinine.

While such a view is formally correct, it misses a deeper understanding of both antibiotics and galenicals. Antibiosis is a phenomenon that was observed in the outer world (i.e., without reference to the human organism). The great achievement of antibiotic research, it has aptly been said, was the concept that the molecular weapons used by bacteria "could be appropriated by man and transferred from the macrocosm to the microcosm."[9] Historically speaking, the pattern of macrocosm-microcosm plays a much greater role in the ancient ideas of cosmic sympathy and antipathy and even more so in the medical philosophy of Paracelsus than it does in Galen's thinking.

On the other hand, the current definitions of galenicals are mere conventions originating from the clash between Galenists and adherents of Paracelsus. In the 1882 edition of Blancard's medical dictionary, we have the following entry: "*Galenica Medicamenta* is the name given to: (1) all simples of whatever kingdome, and (2) all compounds which need only relatively simple preparation, for instance, decoction and infusion. Those compounds which are prepared more laboriously are called spagyric and chemical medicaments."[10] As is to be expected, the editor, Dr. Kühn, an eminent authority on Galen, was right in allowing simples from all three kingdoms—vegetable, animal, and mineral. Galen himself, in the preface to the ninth book of *On the Temperament and Power of Simple Drugs*, wrote: "All that concerns the parts, fruits, juices and liquors of plants has been said before. Now it is our aim to go through the remaining drugs, those produced by mining and the kinds of earth. Afterwards something will also be said about the parts of animals in so far as we use them as drugs for healing." And in the third chapter, dealing with "metallic drugs," he remarked: "Physicians are wont to call metallic drugs those which originate in the mines by spontaneous growth or by means of the furnace. There have to be added as a third group those which men make out of the former in whatever way, for instance, white lead, itch-salve, sandyx, and phykis."[11]

To be sure, Galen by preference used vegetable remedies and was very cautious with respect to mineral drugs, especially their internal use. Yet no sharp line was drawn against chemical preparations and, with the advance of chemistry during the Middle Ages and the Renaissance, new chemical drugs were accepted. Paracelsus's emphasis on the superior nature of chemical preparations, his claim that the chemist should prepare the "arcanum," the potent healing essence in drugs—mineral, vegetable, and animal alike—this together with his advocacy of "poisons" caused the cleavage between Galenists and Paracelsists.[12] Now it was that galenicals in the conventional sense became known in contradistinction to the chemical remedies.[13] Obviously, then, galenicals stood for a medical system, a system we shall now try to analyze with reference to therapy. To this purpose we shall begin with a brief outline of Galenic pharmacodynamics and shall test its implications against later developments.

The crude outlines of Galen's pharmacological system are so well known that we need merely refresh our memory. Its roots can be found in Aristotelian science, the doctrines professed in certain Hippocratic writings, and a vast body of empirical material inherited from previous centuries. From Aristotle, Galen accepted the theory of the four qualities—moist, dry, cold, and warm—constituting all bodies both dead and alive. The mixture of these qualities determines man's "temperament" as a whole as well as the temperaments of the various body parts and organs. Since the four humors of the body correspond to the four combinations of the qualities, their harmonious mixture or disproportion can also be expressed in qualitative terms. A head cold, for instance, is characterized by a superabundance of phlegm, which is cold and moist. It is, therefore, a cold and moist disease, as are all diseases due to phlegm.

This theory of disease enables Galen to find the indication for the treatment of a disease by invoking the famous formula *contraria contrariis curantur*. Disease is cured by bringing the qualitative mixture back to normal. And this is done through remedies whose qualities are opposed to the disease. Our cold would have to be treated by warm and dry remedies, perhaps going to bed and drinking a glass of mulled wine. Galenic pharmacology as a science is to a large extent the finding of the qualities of the so-called simples. How were they ascertained? As an example, let us follow Galen's attempt at elucidating the qualities of vinegar.

Dressings with vinegar, if applied to healthy parts of the body, seem

cold at first, warm afterwards. Consequently, vinegar is cold but also, to some degree, warm. This is important because Galen tried to establish the degrees—four altogether—of intensity to which the qualities were present in the drugs. Thereby it became possible to combine simples so as to achieve a well-calculated total effect. But to return to our example, testing of vinegar in the healthy is not enough; it must also be tested in morbid conditions to find whether it really has a cooling effect.

> When a person is burnt by the application of thapsia upon the skin, vinegar relieves the burning heat. And whoever wishes can learn this by experience, as we did on ourselves in making an exact test of the power of the drug. We anointed our shins in many places [with thapsia] and when, after four or five hours they began to burn and to be inflamed, we sprinkled vinegar on one place, water on another, olive oil on a third; we anointed a fourth with rose oil and others with substances which we believed capable of blunting sharpness or of cooling heat. And the vinegar was found to be more active than any of these.[14]

I hasten to add that I am fully aware of the fallacy of these experiments. The main point is to note for later reference that Galen did experiment and to indicate a main feature of Galenic pharmacology that I trust has become clear, namely, the complementary nature of disease and drug. The nature of the drug is established relative to the human body. The drug is supposed to bring about a change in the body because the disease, too, is a change of the body, an abnormal condition, not an independent entity.

Such a statement has little meaning until some of its consequences become clearer. The easiest way of achieving this is, perhaps, to mention the difference between an apothecary and a physician. It is a familiar occurrence to see a man in a drugstore ask for a remedy against his cough. The apothecary will sell him something directed against this complaint. There was a good deal of the apothecary in Galen, some of whose pharmacological works list prescriptions for various diseases and symptoms.[15] But the physician Galen, as he presents himself in his medical work, knew better. A disease is not only a change of the body but also a change of this particular body. The physician must, therefore, in his treatment take account of the peculiarities of the case—which means that only a physician should treat. We may here quote from a letter Galen wrote to the father of an epileptic child outlining the treatment the boy should follow. A large

part is devoted to the dietary regimen. Finally, Galen mentions a remedy prepared from squills and expresses the hope that the disease unless "very severe and hardened . . . will cede completely in forty days through this drug." He claims to "have cured innumerable children in this way." Now it looks as if Galen, too, were set upon curing a disease, epilepsy, by a specially prepared drug. Therefore, the turn he gives to the matter a little later is all the more interesting. Not only must the ingredients of the prescription be adapted by the attending physicians, but "it is necessary to change and vary everything else which has been mentioned, according to the daily condition of the body, as in all other diseases too." The whole paragraph culminates in the statement: "as all the best physicians have agreed, the drugs are rather the means of assistance than assistance itself."[16]

We can, perhaps, make the Galenic principle of individualized treatment clearer by citing the remarks of two physicians, Leoniceno, possibly a teacher of Paracelsus,[17] and Paracelsus himself concerning the treatment of syphilis. Leoniceno was a Galenist and a humanist. His booklet on the disease is learned as well as steeped in the tradition of finding a variety of treatments that will allow adaptation to the symptoms. He is concerned about the practice—already in vogue—of applying mercury, the alleged specific, indiscriminately. "Let doctors beware," Leoniceno wrote in 1497, "lest after the fashion of a bad cobbler who tried to fit everybody with the same shoe, they also endeavour to cure the French disease in everybody with the same medicine."[18]

Paracelsus's outburst comes about forty years later; it sounds somewhat similar, yet has a different twist. "You know," he tells his medical contemporaries,

> that *Argentum vivum* is nothing but poison and daily experience proves it. Now you have this in use, you anoint patients with it, much more thickly than a cobbler anoints leather with grease. You fumigate with its cinnabar, you wash with its sublimate and do not wish people to say it is poison; yet it is poison and you introduce such poison into man. And you say it is healthful and good, it is corrected with white lead, just as if it were not poison . . . For you know not the correction of Mercury, nor its *Dosin* but you anoint with as much as will go in.[19]

Unless I am mistaken, Paracelsus's indictment is not directed so much against lack of individualization as against ignorance of the true way of

making a poison act as a drug. In modern language, Leoniceno objects to standardization of treatment, while Paracelsus objects to bad and ignorant standardization.

We have chosen examples pertaining to the history of syphilis because of its close connection with the development of specific therapy, chemotherapy, and antibiotic drugs, as indicated by mercury, arsphenamine, and penicillin. Yet not only the treatment but the disease itself has provided the testing ground for Galenic views. Its real or alleged novelty made syphilis more than any other disease appear as an entity and stimulated the development of what we call the ontological concept of disease,[20] clearly and programmatically formulated by Sydenham. In the final analysis, the ontological concept of disease suggests a therapy that is directed against the disease. Sydenham believed diseases to be reactions of the body to occult causes that determined the species of diseases. Insofar as they are reactions, diseases represent an attempt of nature to cure. The physician should support nature in this attempt, and a knowledge of the manifestations of the various diseases will lead to a methodical treatment. But this methodical treatment is not the ideal one. Nature can err, and the morbid reaction, at best, is debilitating. Ideally it should be possible to shortcut the whole process by finding specific remedies that would destroy the disease species directly. According to Sydenham, one such specific remedy was known, namely the cinchona bark that cured intermittent fevers without evacuation of the morbid matter.

> Nevertheless, I have no doubt, but that out of that abundant plenitude of provision for the preservation of all things wherewith Nature burgeons and overflows (and that, under the command of the Great and Most Excellent Creator), provision also has been made for the cure of the more serious diseases which afflict humanity, and that near at hand and in every country. It is to be lamented, indeed, that the nature of plants is not more thoroughly understood by us. In my mind, they bear off the palm from all the rest of the *Materia Medica*. They offer also the most reasonable hopes for the discovery of remedies of the sort in question. The parts of animals are too like those of the human body: minerals are too unlike. That minerals, however, are more energetic in satisfying indications than either of the two other classes of remedies, and that the difference in character is the reason for their doing so, I freely confess. Still they are not specific remedies in the sense and manner explained above.[21]

Sydenham expects the plenitude of nature to have provided in each man's country a specific herb to cure any major disease. Here indeed the galenicals seem to triumph. But as we have found, the conventional concept of galenicals does not necessarily harmonize with Galen's own materia medica. In spite of his veneration for Nature, the pagan Galen was not a pious herbalist who believed that the native plants contained wonderful virtues ordained to heal every man until God had commanded his death. And as far as I can see, the idea of plenitude that, as Professor Lovejoy has shown,[22] played a major role in Western philosophical and scientific thought was of little importance to Galen. It is not our concern to find out where Sydenham received his views. We but ask whether there are any Galenic elements in the assumption of a specific healing virtue directed against the species of the disease.

The question allows us to mention an important pharmacological principle of Galen's, namely his belief in the action of drugs as due to the peculiarity of their whole substance. This principle is quite different from that of the active qualities which we have considered so far; it is irrational, as Galen makes clear in the following words: "For the faculties that are vested in the peculiarity of the whole substance have been shown to lack method and reason and can be known through experience only. For we do not know why this stone [he speaks of the hematite] when brought in contact with a bleeding wound stops the hemorrhage."[23] Both Galen and Dioscorides list many substances the alleged action of which rests on experience only and can be explained no more than the magnetic property of the lodestone—a prototype of all occult forces. Since experience devoid of reason is an easy prey to superstition, many of these substances strike us as superstitions. On the other hand, the belief in the existence of pharmacological properties that would yield their secrets only to experiment was a powerful stimulus. It accounts for the mixture of superstition and experimental daring that we find in the therapeutics of the Middle Ages and Renaissance.

The idea of action by virtue of a body's whole substance was increasingly misused to account for any inexplicable phenomenon until it was ridiculed by Molière in the famous graduation scene of the *Malade Imaginaire.* The candidate is asked for the cause and reason why opium puts people to sleep. And he readily answers:

Quia est in eo
Virtus dormitiva
Cujus est natura
Sensus assoupire.

These lines have been quoted again and again to ridicule a medical science that explained the narcotic power of opium by its narcotic virtue. But it should be noted that the quotation should not be directed against Galen, who ascribed to opium a highly cooling effect that could lead to stupor and death. "Therefore," he added, "it falls upon the physician to use it properly by mixing it with drugs that weaken the power of the cold quality."[24]

This brings us back to the question as to whether Sydenham's postulate for specifics was Galenic. In answer we may say that, although Galen admitted specific actions of drugs, he did not go so far as to put too great a reliance on them. To him the methodical search for the qualitative principles that would change the abnormal condition of the body stood in the foreground, since disease itself was a condition of man, not a reaction to a specific cause.

Sydenham's influence made itself felt at a time when the Galenic system, as a distinct body of medical doctrines, was losing credit. This was due to the overthrow of Aristotelian physics by the new physical and chemical sciences. As soon as matter was no longer seen in terms of qualities, the Galenic system lost its presupposition. In medical therapy this abdication left a void for nearly two hundred years. To be sure, the struggle with the chemists was now over; new drugs were added, and many superstitions were discarded, together with the whole *Dreckapotheke*. Physicians tried to gather experiences as best they could, but for the rest, short-lived or one-sided systems alternated with skepticism.

There are some illuminating data available. The following figures have been calculated for the inmates of the Bamberg Hospital for 1798, when Germany was under the influence of the Brunonian system that classified diseases as sthenic or asthenic. There were 480 patients and the per capita consumption of drugs in grams averaged as follows: opium, 3.6; camphor, 11.7; liquor anodynus Hoffmanni, 30.0; Radix Serpentariae, 7.92; cinchona bark, 31.6; and similar amounts of musk, ethyl acetate, arnica, valeriana, angelica, cinnamon, tinctura ferri tonica, and tinctura chinae composita, besides more than 500 grams of brandy.[25] Some twelve years

later when Brunonianism had lost favor, homeopathy recommended itself
by a minute analysis of the effects of drugs. From the voluminous work
of Dr. Hartlaub, *Systematic Presentation of the Pure Medicinal Effects for the Practical Use of Homeopathic Physicians*, I quote here a list of symptoms relating to
cold sensations on the head together with the drugs supposed to cause them:

> A cold sensation on a small spot on the forehead, as though someone were
> touching him there with a cold thumb—Arnica.
> An icy cold sensation in the upper half of the head when he puts his hat
> on firmly—Valeriana.
> A feeling of cold on a small area around the crown of the head, accompanied by the hair standing on end—Manganum aceticum.
> A feeling of cold on the hair—Acidum phosphoricum.
> Simultaneous sensations of heat and cold on the head, accompanied by
> sensitivity of the hair—Veratrum album.
> Slight shuddering of the hairy part of the head, followed by burning itch
> of the integuments of the head. This itching decreases after scratching but
> only to return with increased intensity—Capsicum.
> The head feels as if it were blown upon by a cool little breeze—Colocynthis.[26]

This was written in 1826. Thirty years later, Trousseau and Pidoux still
admitted that therapeutics and materia medica were "in the chaos of a
transition,"[27] and the existence of therapeutic nihilism testifies to the truth
of this statement, no less than do the various medical sects flourishing in
this country. True, pathology was making great advances. But the anatomical trend that pathology followed was, therapeutically speaking, in the direction of surgery. The developments that were later on to influence pharmacology were of a less spectacular nature.

Beginning with morphine in 1806, several vegetable alkaloids were isolated, the effect of which could be studied experimentally. As a result it
was shown that the galenicals contained substances active in various directions. This in itself was a victory of the Paracelsian idea of extracting
the "arcanum" from the simple. Indirectly and slowly, the increasing hold
that chemistry gained over pharmacy changed its industrial aspect. While
formerly an apothecary had been compelled to be his own chemical manufacturer, the Prussian pharmacopoeia of 1827 allowed him to buy chemicals "which can be purchased genuine from industrial plants and the
preparation of which by the apothecaries is not without some danger and

inconvenience."[28] In 1862 the same apothecary was not even forced any longer to prepare all galenicals. As a result, the galenicals in the conventional sense lost their privilege of native drugs and were increasingly replaced by industrial products. The physician lost his interest in botanical studies; he became dependent on information supplied by manufacturers. While formerly wont to compose an individualized prescription from a relatively small number of drugs, largely simples, he now turned to the prescription of medicines standardized as to dosage and mode of application.

Under these circumstances it is very difficult to affirm or deny connections between Galenic ideas and the new medical treatments that evolved in the late nineteenth century. For instance, the statement that "pharmacology has for its object the recognition and study of all changes which a foreign body can undergo or produce, otherwise than traumatically, in the organism" would agree perfectly well with Galenic principles. It is taken from the *Experimental Pharmacology* by Hermann,[29] who like his more famous French contemporary, Claude Bernard, investigated the action of drugs experimentally. It may be rightly said that these men tried to make pharmacology fit into pathological physiology, thus continuing in Galen's footsteps. But with equal justice one may point out the methodological differences that divide Galenic and modern science and experimentation. Or should we claim antidiphtheria serum or liver therapy to be galenical because Galen, too, used antidotes and animal substances? These are formal analogies which, I am afraid, will not lead us very far.

However, we have not yet completed our survey of Galenism in medicine. Indeed, we omitted one of its most important features, namely the latitude Galen allowed for health. According to Galen, disease is present only when man's functions, biological as well as social, are impaired. In between the ideal disposition of a person in perfect health, on the one hand, and disease, on the other, there exists a wide area extending from relatively good to relatively poor health. The existence of this area allows the Galenic physician to promote health actively, since to him it is not the mere absence of disease.[30] This implies that everybody must constantly watch his health, since it is always prone to deterioration through a wrong manner of life.

At last we have reached a point where we can rightly claim Galen among the moderns. The slogan of "positive health" presupposes a more or less

of health beyond the bare alternative of healthy or sick. Our growing emphasis on preventive medicine, too, begins to resemble the constant concern over health of Greek medicine.

If there is similarity where personal hygiene is concerned, there is also a discrepancy in curative medicine, especially in the branch that, next to surgery, has celebrated the most spectacular triumphs, the cure of infectious diseases. Fracastoro's "seeds" of infection and Sydenham's "species" of disease have become visible in the form of pathogenic micro-organisms and can now be shot by "magic bullets," to use Ehrlich's term in an extended way covering the various miracle drugs.

The aim of using drugs against causes of disease that come from outside is even more foreign to Galenic medicine than treatment of a disease entity instead of the changed body. To be sure, Galen used vermicides and knew of antidotes against poisons. But the undeveloped doctrine of infection points out the difference. The existence of infectious diseases was known, contact shunned, fires lighted to destroy miasmata in the air, and aromatic substances used to counteract them. These were measures of public health and preventive medicine, so to speak, on a par with the avoidance of bad food, unhealthy winds, and so forth. The treatment of infected persons, however, was like that of other patients. The fluent boundary between health and disease was paralleled by a fluent relationship of food and drug. A substance is food as far as it is assimilated by the body but a drug insofar as it alters the body.[31] Galen's materia medica is integrated into his whole system. Dietetic medicine regulates man's life in all its aspects because disease is a deviation from right living. As to "magic bullets," the galenical arrows would never have gone straight to the goal. Even more important than the lack of the weapon was the lack of the intent to cure disease by shooting at the cause.

We have now come back to the modern drugs, including the antibiotics we started with. We raised the question as to the relationship of antibiotics and galenicals and found that the answer depended upon a proper understanding of the term *galenical*. In the quest for such understanding, we were led to an evaluation of some Galenic principles against the background of more recent views. Some of these principles are still valid aims today; others are not. As to our initial question, we can now say that antibiotics are not modeled after the galenicals. How far they fit into another historical pattern may become clearer in some other context.

Our answer will hardly cause surprise and, since galenicals and Galenism are not in favor, it will not arouse resentment either. On the contrary, we may be asked why such a long detour was made to reach the obvious. Let us answer this question by raising another one. Why has Galenism in medicine become a word of reproof? Because, I imagine it to be said, it was a system full of errors that subjugated generations into blind adherence until medicine finally got rid of it.

I doubt that medicine has gotten rid of it. To be sure, the system was changed and reformed part by part until it lost all resemblance to the original. Vesalius reformed anatomy, but it was Galen who had made anatomy an integral part of medicine. Harvey reformed physiology, but it was Galen who had bequeathed experimentation to medicine. The iatrochemists and iatromechanists eliminated Aristotelian physics from medicine, but it was Galen who had integrated into medicine what went for physics and chemistry in his time. I need not repeat what happened to Galenic pharmacology nor elaborate the fact that new ideas were added to medicine that were foreign to Galen. In short, since the Renaissance, medicine has, step by step, renovated its house and built new wings, tearing old ones down. But the house itself is still there; it has never been razed, though hardly any of the old bricks remain. The memory of this age-old fight accounts for our enmity as well as for our illusion that Galenism is dead. What is Galenism? Galen himself again and again stated it as his fundamental conviction that the two legs of medicine were reason and experience.[32] We may reason and experience differently—but shall we deny the validity of his conviction? In this sense, and in this sense only, we are still Galenists. To test modern beliefs and practices against Galenism, therefore, means bringing out their position within the development of scientific medicine.

NOTES

1. This is an ad hoc definition from the merely therapeutic point of view of the present chapter. As Professor Waksman rightly reminds us, a general definition of antibiotics must not be restricted to the effect upon pathogenic organisms. Moreover, as Florey stated in 1949, "there is an increasing tendency to enlarge the scope of the word to include antimicrobial substances derived not only from microbes but from any living source, including plant and even animal tissues." H. W. Florey et al., *Antibiotics* (Oxford: Oxford Univ. Press, 1949), 1:15.

2. Ibid., 14 f.

3. Ibid., 1 ff. This statement refers only to the historical quest for therapeutically active antibiotics, for which the article by A. G. Cranch, "Early use of penicillin," *JAMA*, 123 (1943): 123 is an example. The statement does not refer to general investigations of microbial antagonisms carried on before the impact of penicillin; cf. Florey, *Antibiotics*, and Selman A. Waksman, *My Life with the Microbes* (New York: Simon and Schuster, 1954), 208 ff. See also the articles by various authors in "Antibiotics in use," *J Hist Med Allied Sci* 6 (1951).

4. Robert T. Gunther, ed., *The Greek Herbal of Dioscorides*, English ed. by John Goodyer, A.D. 1655 (Oxford: Oxford Univ. Press, 1934), 112. J. Berendes, *Des Pedanios Dioskurides aus Anazarbos Arzneimittellehre in fünf Büchern* (Stuttgart: Ferdinand Enke, 1902), 104, interprets this passage as referring either to rotting wood or, not unlikely, the micro-organisms scooped from the wood.

5. Galen, *De simplicium medicamentorum temperamentis ac facultatibus*, bk. 8, chap. 18, ed. Kühn, 12: 118. The reference to the elm tree may be due to the fact that in Dioscorides the paragraph on this tree precedes the one on rotten wood.

6. Petrus Andreas Matthiolus, *Commentari in Sex Libros Dioscoridis Anazarbei De Medica Materia* (Venice, 1570), 123.

7. Galen, *De simplicium*, bk. 6, chap. 6, 4, ed. Kühn, 11: 882. The "digestive power of yeast probably refers to deep-seated purulent processes. On the use of yeast in antiquity, cf. Florey, *Antibiotics*, 56 f.

8. The definition of *galenicals* is not firmly established. *Stedman's Medical Dictionary*, 17th ed. (Baltimore: Williams and Wilkins), 469, gives three different definitions.

9. William Thomas Salter, *A Textbook of Pharmacology* (Philadelphia: Saunders, 1952), 1056.

10. Stephan Blancard, *Lexicon medicum*, ed. Kühn (Leipzig: Schwickert, 1832), 1:667.

11. Galen, *De simplicium*, 12:159.

12. Robert Multhauf, "Medical chemistry and 'the Paracelsians,'" *Bull Hist Med* 28 (1954): 101–26.

13. George Urdang drew attention to the confused situation resulting from the difference in meaning between the Paracelsian and the modern use of the term *chemical processes* in "How chemicals entered the official pharmacopoeias," *Arch Int Hist Sci* 7 (1954): 303–14.

14. Galen, *De Simplicium*, bk. 1, chap. 21, 11:418 f.

15. See esp. Galen, *De compositione medicamentorum secundum locos* and *De compositione medicamentorum per genera.*

16. Owsei Temkin, "Galen's 'Advice for an epileptic boy,'" *Bull Hist Med* 2 (1934): 179–89, quotes on 188–89.

17. Leoniceno was professor at Ferrara at the time when Paracelsus is supposed to have received his doctor's degree at that university.

18. Owsei Temkin, "Therapeutic trends and the treatment of syphilis before 1900," *Bull Hist Med* 29 (1955): 309–16.

19. Paracelsus, "Seven defensiones," in *Four Treatises of Theophrastus von Hohenheim Called Paracelsus,* ed. Henry E. Sigerist (Baltimore: Johns Hopkins Press, 1941), 22.

20. R. Schneider, *Die Syphilis und die Anfänge der nosologischen Krankheisauffassung,* diss. Leipzig, 1929.

21. *The Works of Thomas Sydenham, M.D.,* trans. R. G. Latham, 2 vols. (London: Sydenham Society, 1848), 1:22.

22. Arthur O. Lovejoy, *The Great Chain of Being* (Cambridge: Harvard University Press, 1936).

23. Galen, *De simplicium,* bk. 9, chap. 2, 12:192.

24. Ibid., bk. 7, chap. 12, 12:73.

25. C. Binz, "Zur Geschichte der Pharmakologie in Deutschland," in *Klinisches Jahrbuch* (Berlin, 1890), 2:3–74, see 58 f.

26. Carl Georg Christian Hartlaub, *Systematische Darstellung der reinen Arzneiwirkungen zum practischen Gebrauch für homöopatische Aerzte,* Ersther Theil (Leipzig: Baumgärtner, 1826), 363.

27. A. Trousseau and H. Pidoux. *Traité de thérapeutique et de matière médicale* (Paris: Béchet Jeune, 1855), CV.

28. Edward Kremers and George Urdang, *History of Pharmacy* (Philadelphia: Lippincott, 1940), 113.

29. L. Hermann, *Experimental Pharmacology,* trans. Robert Meade Smith (Philadelphia: Lea's Son and Co., 1883), 14.

30. Owsei Temkin, in *The Epidemiology of Health,* ed. Iago Galdston (New York: Health Education Council, 1953), 9 f.

31. See Galen, *De temperamentis,* esp. bk. 3, chap. 4.

32. Galen, *Logos and empeiria.* See Iwan von Müller, *Ueber Galens Werk vom wissenschaftlichen Beweis* (München, 1895) (Abhandlungen d. K. bayer. Akademie der Wiss: I. Cl., XX. Bd., II. Abth.), 436, 461.

CHAPTER 10

Nutrition from Classical Antiquity
to the Baroque

Nutrition and medical ideas on nutrition cannot be independent of the style of living. Eating and drinking are part of the civilization in which the physician has to think and act, and this civilization is encompassed within the borders of the known world.[1] Certain foods are prescribed; the cooking and the quantity can be given in detail. But the physician cannot offer that which does not exist or is not available. What is the good of theorizing about all possible foods, when most of them are out of reach? The very fact that the medical literature of the period with which we are concerned tends to cover a large variety of foods indicates the status of the doctor as the attendant of people not in constant want. This literature does not always tell us what those people ate whose diet was strictly limited by necessity.

We all have impressions of the dietary style of some of the periods we are covering. In the Homeric epic poetry, feasting means heavy eating. Penelope's suitors waste Odysseus's estate by their sumptuous meals consisting mainly of roast beef, mutton, goat meat, and pork. If this had been the daily food of the average Greek of the time, Homer would hardly have dwelled on it at length. The staple diet of the average Greek was *maza*, a dried dough made of barley. In early times, bread, particularly wheaten bread, was not common. Of vegetables there existed mainly beans, lentils and peas, onions, leek, and cabbage; cucumbers, olives, and figs were consumed in considerable quantities. There was milk from cows and goats, and cheeses, and, of course, there were grapes and wine, the latter usually

Originally published in *Human Nutrition Historic and Scientific*, edited by Iago Galdston (New York: International Universities Press for the New York Academy of Medicine, 1960).

mixed with water. It is not likely that the average Greek ate much meat; pork probably held the first rank. But there the sea offered its supply of food—pickled, salted, or smoked fish was very popular.

The increased trade of Greece facilitated the import of foods from different parts of the world and, more important perhaps, it lessened the dependence on the local supply. Rome, at the height of its power, had to feed the mass of its population with imported grain. Delay in the arrival of the boats or interrupted communications on land could spell famine.

Famine, however, was a more pronounced feature of the medieval inland economy. Famines caused by weather conditions, epidemics, or wars and sieges occurred again and again until far into modern times. Such famines might occasionally affect both poor and rich, but the peasant was more frequently threatened, since he could count on little else than the resources of his immediate neighborhood. The condition of the peasants varied considerably with the times and with individual countries. It ranged from a bread-and-water existence to a comfortable life with ample bread, vegetables, fruits, milk, cheese, meats, and ale. But there was an insecurity threatening the dietary necessities and, consequently, the possibility of "conspicuous consumption" in the form of food and drink. Spices, which had to be imported, were in great demand. The desire to obtain what was rare probably helped to mold taste. The contrast between workday and feastday, between poor and rich, expressed itself largely through the gullet.[2]

Thus, we have pictures of changing nutritional styles. The Greeks of the classical period and the Romans of the early Republic appear to us as moderate in their habits. The very name of Sybarite, singling out the inhabitants of the city of Sybaris with its fertile land, indicates that voluptuous living was not considered common. The proletarian of imperial Rome demands bread and plays; his rich and degenerate fellow citizen titillates his palate with a feather to relieve his stomach for more food. This picture gives way to Breughel's vomiting peasant who has gorged himself at a feast: In the baroque period of the seventeenth and early eighteenth centuries, we find in juxtaposition starvation and scurvy and high living and gout. In between, from the Middle Ages on, is the religious ascetic whose privation is self-imposed. In the Renaissance there appears at his side the Erasmus-like humanist, who not only preaches moderation but whose enjoyment of intellectual values frees him from the excessive pleas-

ures of the stomach. Toward the end of this period, changes took place which again were to modify the picture. In England, the industrial revolution posed new problems, and the potato, cultivated gradually from the seventeenth century on, offered a new indispensable food. Tea and coffee came into vogue, first in the elegant world, then in a broader stratum.

The dependence of the medical man on the nutritional style of his time is obvious. Not only does it determine the prevalence of certain diseases; it also determines what the physician is supposed to know. The ancient and medieval doctor knew nothing of vitamins. But he knew something about the differences between the various dishes that were served; for instance, the gobio, a fish of the gudgeon kind, was evaluated differently according to its provenience.[3] To cite a further example, in 1766, Pennier de Longchamp, a French physician, published a dissertation on truffles and mushrooms. His curiosity had been aroused because at a party a young man was observed indulging heavily in truffles. "Since my profession obliges me to know the virtues of all the plants, I was addressed and asked what effects truffles had."[4]

But in spite of prevalent styles and fashions, there also existed a form of nutritional wisdom that withstood changes. It is documented in the *Regimen sanitatis Salernitanum,* a Latin poem that purports to speak in the name of Salerno to the king of England. This poem certainly existed by the end of the thirteenth century. Its dependence on traditional medicine is obvious, yet its connection with Salerno has rightly been doubted. It gives the impression of a popular dietetic work written by a medical man for the educated reader (i.e., whoever knew Latin) in a tone devoid of scholastic ponderosity. It stands between antiquity and modern times. Starting with this poem, we shall then look backward and forward. "If you lack doctors," we read, "these three will be doctors for you: a cheerful mind, rest, and a moderate diet." The idea of moderate eating is repeatedly emphasized: "A large supper imposes the greatest punishment on the stomach,"[5] and one should not eat before the previous meal has been digested, for which appetite is a sure sign. There are also, at the very beginning, summary rules for what should or should not be eaten:

> Peaches, apples, pears, milk, cheese, and salted meat,
> Deer, hare, goat, and veal,
> These engender black bile and are enemies of the sick.

Fresh eggs, red wines, fat broth,
Together with fine pure flour, strengthen nature.
Wheat, milk, young cheese,
Testicles, pork, brain, marrow,
Sweet wine, food pleasant to the taste,
Soft-boiled eggs, ripe figs, and fresh grapes,
Nourish as well as make fat.[6]

In reading these lines the older among us may be reminded of the days when the doctor was not allowed to leave without having discussed the diet of his patient. My parents' faith in the physician did not necessarily go as far as to order the prescription he left, but I was sure to get chicken broth or cream of wheat if he so advised. The lines also indicate the audience who might benefit by them. When I said that the poem was for the educated reader, I should have added that it was also for the well-to-do reader. The dietary wine list alone bears that out: Wines should be clear, aged, subtle, mature; sweet white wines are considered particularly nourishing. Nevertheless, there are also things for those of modest means: "Cheese and bread are good food for those in good health—if a man is not healthy do not add the cheese to the bread."[7]

Before leaving Salerno, I would like to quote two lines that seem to sum up the implications of the practical side of medical nutrition:

> *Quale, quid, et quando, quantum, quoties, ubi dando,*
> *Ista notare cibo debet medicus diaetando.*[8]

The quality and the kind of food, the when, and how much, and how often, and where to be given—these are the things the doctor has to point out in the diet. There is only one question left out, viz., *quare?* why? It is the why of the regulations that concerns the doctor and leads to a science of nutrition.

A medical science of nutrition presupposes that man's food should be regulated to keep him free from disease and cure him of sickness. We are aware of the fact that Greek medicine held this belief, but we must now probe its strength and possible roots. Immediately we are confronted with the phenomenon that, for many Greek physicians since the Hippocratic era, scientific medicine was identical with dietetic medicine.[9] The concept of diet was broader than it is today.[10] It comprised not only food and

drink but also work, sleep, climate of the home, emotions, and sexual life—that is, what the medieval doctors came to call the six *res nonnaturales,* the six "nonnaturals."[11] Since food and drink occupied the main position in the regimen, we shall use the term *diet* in this restricted sense.

To the author of the Hippocratic work *On Ancient Medicine,* medicine emerged from the observation that sick people could not stand the same food as the healthy. To him, this was an elaboration of a much wider thesis: that the diet of civilized man owed its origin to the hardships to which primitive people were subject, people who, like animals, lived on the raw products of the land.

> As it is, sheer necessity has caused men to search for and discover medicine, because the same diet did not benefit, just as it does not benefit today, sick and well alike. To go further back still; I hold that the regimen and nutriment which men in health now use would not have been discovered at all, if men were satisfied with the same food and drink as satisfy an ox or a horse or any creature except man, for example, the things that grow out of the earth—fruits, wood and grass.[12]

Apart from its wide sweep, this treatise is remarkable for basing nutrition on observation. This, too, was taken up by later authors. Galen applauded Diocles of Carystus, who, in the following century, warned against too great a faith in one's ability to give the cause of every effect: it was not always possible to say why foods were nourishing, or diuretic, or acted as a cathartic.[13] As a contrasting example of causal nutrition, I quote this passage from the Hippocratic *Regimen in Health:*

> In summer the barley-cake [the maza] is to be soft, the drink diluted and copious, and the meats in all cases boiled. For one must use these, in the summer, that the body may become cold and soft. For the season is hot and dry, and makes bodies burning and parched. Accordingly these conditions must be counteracted by the way of living.[14]

If medicine did not originate directly from nutritional observations, there is yet little doubt that dietetic medicine, rather than surgery or pharmacology, gave the main impetus to speculative reasoning.[15] Why, then, did nutrition attain this great significance in classical antiquity, probably in the fifth century B.C.? There may have been a variety of reasons; together

they amount to the belief on the part of the physician that he could form the body of man, just as the contemporary sophist believed that he could form his mind. Surgery dealt with injuries and external disorders. The use of drugs had been established by tradition. But in prescribing food and drink, the physician had the example of the gymnast, the trainer, who built up his protégé to the necessary strength. The preoccupation with the athlete that the Hippocratic doctor shared with the gymnast is indicated in the third Aphorism, which begins: "In athletes, a perfect condition that is at its highest pitch is treacherous. Such conditions cannot remain the same, or be at rest."[16]

Nutrition was not only important in the treatment of disease; it guided the life of the healthy as well. It is but another side of the same picture if Greek physicians found the nature of disease in dietetic errors. According to Aristotle, Hippocrates explained diseases in the following way:

Either because of the quantity of things taken, or through their diversity, or because the things taken happen to be strong and difficult of digestion, residues are thereby produced, and when the things that have been taken are too many, the heat that produces digestion is overpowered by the multitude of foods and does not effect digestion. And because digestion is hindered residues are formed. And when the things that have been taken are of many kinds, they quarrel with one another in the belly, and because of the quarrel there is a change into residues. When however they are very coarse and hard to digest, there occurs hindrance of digestion because they are hard to assimilate, and so a change to residues takes place. From the residues rise gases, which having arisen bring on diseases.[17]

The rest of the passage, which is taken from Jones's translation of the so-called *Anonymus Londinensis,* need not concern us here, nor need we enter upon the hotly debated question as to whether this really was the opinion of the great Hippocrates. What Jones's translation rightly calls "digestion," if rendered literally, means cooking, and this cooking process, we read, is sustained by the heat of the body. Digestion as cooking was generally believed to be a Hippocratic idea; it was shared by Aristotle and Galen; it remained the favored theory throughout the Middle Ages and the Renaissance. Together with it went the view of food and air as fuel for the innate heat. But the right kind of food was needed; too much food or wrong food can suppress the fire. I take the following quotation from Galen:

Just as the proper nutriment for fire increases it, in the same manner nutriment that is proper and natural for bodies warm by nature, will always strengthen them and increase their inborn heat. And this is the property of all food. Sometimes, however, too much food weakens the innate heat, after the manner of wet wood. For this too is nutriment for the fire, but if heaped on the fire it is too much and therefore conceals the flame and threatens extinction.[18]

The inborn heat digests food and is, in turn, maintained by food. Food is needed to replace the losses of the body, especially those incurred by insensible transpiration.

The quantity of food must be proportionate to the quantity of transpiration. If nothing in us flowed away, we would not have come to need food, since the original mass of [body] substance would have been preserved for ever. However, the animal bodies are dissipated by the transpiration that takes place through the pores invisible to the sense. Therefore we need food. For nutrition is the restitution of what has been emptied.[19]

Here Galen gives us a fundamental definition of nutrition. Food, including air, is needed because the body is constantly discharging material in the process of transpiration: waste products that leave it through the invisible channels of the skin.[20] Here is the idea of nutrition as replacement in a process of perpetual combustion. Put this way, Greek physiology does not have the static appearance with which it is sometimes credited. There is a constant coming and going of food, drink, and air to feed the body heat and its regulator, the vital pneuma, and to make up for the waste.

But, we are reminded, these ideas lack quantitative reasoning and are thereby sharply at variance with the modern approach. Strictly speaking, this is not even true as far as Galen is concerned. Nutrition to him is that which replaces the material losses of the body without changing it. Herein lies the basic difference between nutriment and drug. What is assimilated and thus helps the body to grow or to maintain its status is nutriment. That which changes the temperament of the body is drug or poison.[21] Just as Galen has a quantitative scheme for the potency of drugs, so for food, too, he has a quantitative scheme. If a person's temperament deviates to one side, being too warm by three (imaginary) units, the food must deviate to the same degree in the opposite direction (i.e., it must be three units

too cold).[22] Then, and then only, will it have a purely nutritional effect without increasing the abnormal deviation.

Admittedly, these units (*arithmoi*) of which Galen speaks are purely imaginary and cannot be measured. But four hundred years before Galen, it had occurred to Erasistratus to measure emanations from the body.

> If one were to take a creature, such as a bird or something of the sort, and were to place it in a pot for some time without giving it any food, and then were to weigh it with the excrement that visibly has been passed, he will find that there has been a great loss of weight, plainly because, perceptible only to the reason, a copious emanation has taken place.[23]

If in this quotation we replace the bird with the scientist experimenting on himself, we have the experimental design of Santorio Santorio depicted in his *Ars de statica medicina* of 1614. Erasistratus speaks of "invisible emanation" while Santorio attributes the loss in weight to "insensible perspiration."

> If eight Pounds of Meat and Drink are taken in one Day, the Quantity that usually goes off by Insensible Perspiration in that Time, is five Pounds.[24]

Such losses must, of course, be replaced if man is to survive. The measurement of insensible perspiration is, therefore, a means of regulating the nutritional requirements, a regulation where the physical properties of food are decisive. "The most Liquid Parts of our Food are likewise the most heavy, and the Solid lighter: Bread and Flesh are light, Wine and Broaths heavy. A Glass of Wine is almost three Times as heavy as a Piece of Bread of the same Bulk."[25] *Heavy* and *light* here refer to the specific weight. Moreover, regulation is safer than reliance on the instinctive demands of the healthy body—an opinion referred to the authority of Celsus.[26]

The obvious errors of Santorio—his attributing to perspiration what is largely a matter of respiration—must not prevent us from seeing the significance of his vision. Nutrition is viewed as an isolated problem that can be solved on an exact basis of measurement. Santorio preceded Harvey, who more felicitously solved the problem of the motion of the heart and blood as a circumscribed problem of cardiovascular physiology.[27] We shall see that a similar step was made on the biochemical side.

Santorio's ideas have been compared with the system of bookkeeping

then in vogue among Italian men of business.[28] It is quite possible that this exerted an influence on him, just as the influence of his contemporary, Galileo, cannot be doubted. However, the existence of a similar thought, some nineteen hundred years before in Alexandria, should make us hesitate to take the analogy between bookkeeping and metabolic balance as proof of influence.

In Santorio's scheme of physiology, food whose particles could easily pass through the ducts of the body and be "perspired" was less nourishing than food of gross particles. Borelli, one of the greatest of the baroque biophysicists, declared that digestion was a grinding of the food into minute particles which, circulating with the blood, could be used for new building material. Erasistratus, too, had defined digestion as a grinding process—an interesting example of the consequences to which similar basic presuppositions will lead across the centuries and differences of so-called intellectual climate.[29]

I am comparing ancient ideas with those of the sixteenth and seventeenth century, well aware of the perils of such a summary procedure. The Middle Ages are a period that cannot be dismissed with a reference to the School of Salerno. Especially now, when we turn to the chemical aspects, we cannot overlook entirely the role of alchemy, ancient and medieval. One example may suffice.

There is a chain of ancient philosophers and physicians, beginning with Empedocles, who declared digestion to be a process of putrefaction.[30] The fragmentary nature of the material outside the Galenic tradition does not allow us to determine what exactly was meant by this definition. It may have been the obvious change from fresh food to feces. It may also have implied a closer comparison with the decay of vegetal matter such as stated by an author of the eighteenth century:

> Vegetable Putrefaction resembles very much Animal Digestion. Vegetable Putrefaction is produced by throwing green succulent Vegetables in a Heap in open warm Air, and pressing them together, by which all Vegetables acquire, First, A Heat equal to that of a Human Body. Secondly, A putrid stercoraceous Taste and Odour, in Taste resembling putrid Flesh, and in Smell Human Foeces.[31]

But then we also have the symbolic language of the alchemist. By late antiquity, alchemical processes were compared with the growth of the em-

bryo and the nourishment of the body. Putrefaction in late alchemistic parlance means revival, the transformation of a substance through its death.[32] On the other hand, the Arab physician and alchemist Rhazes not only thinks of the plague as leading to putrefaction of the blood but compares this putrefaction to hot, fermenting must.[33] The ideas of putrefaction and fermentation were never quite apart. Van Helmont speaks of digestion as a kind of putrefaction, and he thinks of it in terms of fermentation.[34] What seems new is the emphasis, rather than the details, which, as the comparison of the mechanists of the third century B.C. with those of the baroque showed, can be found a long way back.

Paracelsus claims that all food contains nourishment and poison. This may not seem too far removed from Galen's opinion that most foods also act as drugs. But then Paracelsus says that each creature "has within his body the agent that eliminates this poison from what the body takes up. This is called the alchemist because it uses the art of alchemy."[35] If the alchemist fails, disease will result. It is easy to see how such concepts are in line with a chemical interpretation of normal and pathological metabolism. The Galenic idea of the inborn heat, as Pagel has shown, meets opposition even before Paracelsus. Reuchlin, the humanist, argues that if heat were responsible for gastric digestion, heat outside the body should effect this even better.[36] With van Helmont, the heat of the body is but an adjunct of life, not a primary fundament.[37] The demotion of body heat from a vital principle to a vital concomitant of life is one of the premises of Lavoisier's work. To claim, as Lavoisier did, that warmth is communicated to the body through the blood as a result of chemical combustion in the lungs, depends on the possibility of viewing it as a separate process.

Van Helmont distinguished six successive forms of digestion, beginning in the stomach and ending in the tissues where assimilation was completed. As Michael Foster has shown, van Helmont was a careful observer. More recently, Multhauf and Pagel debated van Helmont's relationship to older, even Galenic ideas of fermentation and the use he made of his knowledge of acids.[38] Pagel has brought forward arguments to the effect that van Helmont discovered the presence of hydrochloric acid in the stomach.[39] There is agreement that van Helmont was a vitalist and that his chemistry cannot be altogether separated from his cosmic speculations. But his successors, the iatrochemists, from the middle of the seventeenth century, dissociated themselves from his philosophy. They stood close to

Descartes and Boyle, being mechanists on the molecular level.[40]

With these iatrochemists there began a development on the chemical side that culminated in Lavoisier's discovery of respiration as an oxidative process. Carbon dioxide was discovered by van Helmont. The remainder of the period is divided into two phases. The first is characterized by the names of Boyle, Hooke, and Mayow. Dr. Guerlac's investigations are changing our opinions as to the meaning of some of these and later contributions.[41] I shall limit myself to a mere reminder of what seems fairly well established about the second phase, which begins with Stahl's phlogiston theory.[42] This theory claims that, in combustion, something (viz., the phlogiston) leaves the burning substance. Without challenging this wrong assumption, the chemists of the eighteenth century arrived at a better knowledge of the gases needed for or involved in combustion. Carbon dioxide, van Helmont's "gas sylvestre," is studied by Joseph Black as "fixed air."[43] He finds that it is contained in the exhalations of animals. A few years later, Henry Cavendish studied "inflammable air" (i.e., hydrogen) and Daniel Rutherford discovered "residual air" (i.e., nitrogen), in which neither flame nor animal can exist. With the discovery of oxygen by Priestly and Scheele in the early 1770s and Crawford's measurements of heat production in man, it needed only Lavoisier's "brilliant mind to mold these miscellaneous achievements into a harmonious entity."[44]

It is not my intention to describe in detail the work of Lavoisier and Laplace.[45] I am concerned with the road leading to their discoveries. Here, then, the question may be permissible as to why the work of these two— a chemist and a mathematician—marks an epoch in the history of nutrition. One of the answers is that their work made possible a calculation of food requirements in terms of calories and thus an exact science of nutrition.

Now it seems to me that the predecessors of Lavoisier and Laplace can be divided into two groups: the chemists, whom we just discussed and whose solid work formed a basis for advance in the right direction; and those who made it possible to approach nutrition rationally. This leads us back to the iatrochemists and iatrophysicists.

Van Helmont's separation of digestion from the inborn heat led to the same consequences as Santorio's study of metabolism. Both were steps in the dissolution of what we call "life" into a number of physiological functions. Such a dissolution leads to an embarrassing question: How do

all these functions or processes combine to make an individual? This integration could be attributed to the soul, or one could—then as now—look at these processes with the eyes of the engineer, intent on the maintenance of a machine rather than on understanding. Thus in 1730, John Quincy, in an English edition of Santorio's work, spoke ironically of van Helmont. However, he also said:

> In the whole then, so far as a Person can consider a humane Body as a Machine, and by the known Laws of Motion demonstrate the Powers and Operations of its several Parts, so far may be with certainty known how to manage it, in order to produce any Change therein; that is, if he has proper Instruments, and upon the same Principles understands their Efficacy and Manner of Application.[46]

In the early eighteenth century both physics and chemistry suggested their methods to anybody wishing to base his diet on scientific principles. On the chemical side, John Arbuthnot offered his popular *Essay Concerning the Nature of Aliments,* from which I cite the following example concerning the meat of fish, which he thought to be more watery than that of terrestrial animals. "From which Qualities," he writes, "a Diet of Fish is more rich and alkalescent than that of Flesh, and therefore very improper for such as practise Mortification. The Inhabitants of Sea-Port Towns are generally prolifick."[47]

Arbuthnot wrote his book with the express intent of showing that nutrition, too, had a scientific basis. Whether he, John Quincy, and others succeeded in convincing the world of this truth is hard to tell. That all nutritional theories before Lavoisier contained shortcomings is no counterargument. People, including physicians, may believe what a later generation deems wrong. Surely the qualities of the ancients are not more convincing to us than Santorio's scales and the acids and alkalis of Arbuthnot. Yet ancient medicine put its trust in nutrition. Nevertheless, I am inclined to say that medicine at the end of the baroque period was no longer dietetic medicine. True, the old tradition was still strong; physicians told their patients what to eat and drink, and they preached temperance to those afflicted with gout. Moreover, palpable progress was made in the use of oranges, lemons, and limes for scurvy. But it must be remembered that when we think of Paracelsus and the iatrochemists, we think of new drugs and the apothecary's shop. When we think of Syden-

ham, we think of the concept that diseases are entities prevalent at certain times rather than consequences of dietary mistakes. When we think of Ramazzini, we think of diseases as caused by one's occupation rather than by food. In short, the role of food is not denied, but neither does it stand in the foreground of attention. Why this change in comparison with antiquity?

Dietetic medicine was based on the idea that causation of health and disease was in man's power. If diseases, short of epidemics, were caused by dietary mistakes, then knowledge and character determined health and disease. Such a medical system was also based on the view that the physician could build and rebuild the human body and, with the body, perhaps even the mind.

In 1747 Gaubius,[48] wistfully and with regret of past glory, quoted these words of Galen:

> Let those who refuse to admit the efficacy of food in making men better or more dissolute, more unrestrained or more reserved, bolder or more timid, more barbarous or more civilized, or more given to disputes and fighting, let them on thinking better of it inquire in order to learn from me what they should eat or drink. For they will profit mightily with regard to moral philosophy, and furthermore they will enhance the virtues of the logical soul, becoming more intelligent, more studious, more prudent, and acquiring a better memory. In fact, I shall instruct them not only as to nourishment, as to what to drink, and concerning the winds, but also concerning the temperaments of the surrounding air, and I shall teach them also what regions they should seek out, or flee.[49]

At the time of Lavoisier, such faith in dietetic medicine was gone. But was the century without any faith in nutrition? In answer, I quote John Quincy's statement: "I am not at all unaware how severe some will be hereupon, in requiring how often they must weigh themselves, and whether they ought to eat and drink by the Ounce."[50] What a modern ring this has! Significantly enough, this remark expresses the sentiments of people at large rather than those of the doctor. If I interpret it rightly, it means the anticipation of a science of nutrition in which authority has switched to measurements and figures. But around 1750 this was no more than a foreboding. In the meantime, people enjoyed a nutritional laissez faire in which neither the imperious doctor nor cold facts tyrannized them.

NOTES

1. Henry E. Sigerist, *Civilization and Disease* (Ithaca: Cornell Univ. Press, 1943), 14.

2. K. Hintze, *Geographie und Geschichte der Ernährung* (Leipzig: Thieme, 1934).

3. Galen, *Opera omnia*, ed. C. G. Kühn (Leipzig: Cnobloch, 1821–33), 6:718.

4. Henry E. Sigerist, "American truffles: a Thanksgiving fantasia," *Bull Hist Med* 16 (1944): 402–9, see 403.

5. Francis R. Packard, ed., *The School of Salernum* (New York: Hoeber, 1920), 159, 160.

6. Ibid., 160.

7. Ibid., 164.

8. Ibid., 168.

9. Owsei Temkin, "Greek medicine as science and craft," *Isis* 44 (1953): 213–26.

10. Ludwig Edelstein, "Antike Diätetik," *Antike* 7 (1931): 255–70.

11. E. T. Withington, *Medical History* (London: Scientific Press, 1894), 389.

12. W. H. S. Jones, *Philosophy and Medicine in Ancient Greece* (Baltimore: Johns Hopkins Press, 1946), 67.

13. Galen, *Opera omnia*, 6:455 f.

14. W. H. S. Jones, trans., *Hippocrates* (London: W. Heinemann, 1923–31), 4:45–47.

15. Edelstein, "Antike Diätetik"; Temkin, "Greek medicine."

16. Jones, *Hippocrates*, 4:99.

17. W. H. S. Jones, *The Medical Writings of Anonymus Londinensis* (Cambridge: Cambridge Univ. Press, 1947), 35–37.

18. Galen, *Opera omnia*, 15:265 f.

19. Ibid., 15:240.

20. The Greek word is *diapnoē*, which may mean respiration as well as transpiration, including not only gases but sweat as well.

21. Galen, *Opera omnia*, 6:468.

22. Ibid., 6:469.

23. Jones, *Medical Writings*, 127.

24. John Quincy, trans., *Medicina Statica: Being the Aphorisms of Sanctorius*, 2nd ed. (London, 1720), 47.

25. Ibid., 58.

26. Ibid., 188.

27. Henry E. Sigerist, *The Great Doctors* (New York: Norton, 1933), 143, 154 f.

28. Johann Daniel Achelis, "Die Ernährungsphysiologie des 17. Jahrhunderts," *Sitzungsberichte Heidelberger Akad Wissenschaften* 3 (1938): 9.

29. Alphonsus Borelli, *De Motu Animalium* (Rome: Angelus Barbo, 1681), 394; Aurelius Cornelius Celsus, *De Medicina* (Florence: Nicolaus Laurentius, 1478).

30. Max Wellman, *Die Fragmente der sikelischen Ärzte* (Berlin: Weidmann, 1901), 34.

31. John Arbuthnot, *An Essay Concerning the Nature of Aliments* (London: Tonson, 1732), 10.

32. Gino Testi, *Dizionario di Alchimia e di Chimia Antiquaria* (Rome: Casa Editrice Mediterranea, 1950), 148 f.

33. Walter Pagel, *Paracelsus* (New York: Karger, 1958), 174.

34. Ioann Baptista van Helmont, *Ortus Medicinae* (Amsterdam: Elzevir, 1652), 157.

35. Bombastus von Hohenheim, *Volumen Medicinae Paramirum*, trans. F. Leidecker (Baltimore: Johns Hopkins Press, 1949), 29.

36. Pagel, *Paracelsus*, 293.

37. van Helmont, *Ortus Medicinae*, 160; K. E. Rothschuh, *Geschichte der Physiologie* (Berlin: Springer, 1953), 47.

38. M. Foster, *Lectures on the History of Physiology* (Cambridge: Cambridge Univ. Press, 1901), 127 ff.; Robert P. Multhauf, "J. B. van Helmont's reformation of the Galenic doctrine of digestion," *Bull Hist Med* 29 (1955): 154–63; Walter Pagel, "J. B. van Helmont's reformation of the Galenic doctrine of digestion—and Paracelsus," *Bull Hist Med* 29 (1955): 563–68.

39. Walter Pagel, "Van Helmont's ideas of gastric digestion and the gastric acid," *Bull Hist Med* 30 (1956): 524–36.

40. Paul Diepgen, *Geschichte der Medizin* (Berlin: De Gruyter, 1949), 1:295.

41. Henry Guerlac, "Joseph Black and fixed air," *Isis* 48 (1957): 124–51, 433–56.

42. Graham Lusk, *Nutrition* (New York: Hoeber, 1933), 37 ff.

43. Guerlac, "Joseph Black."

44. Lusk, *Nutrition*, 51.

45. Lavoisier and Laplace, by means of an ice calorimeter, measured the heat produced by an animal and the heat produced by the combustion of carbon, and they compared these figures with the respective amounts of oxygen consumed and carbon dioxide produced.

46. Quincy, *Medicina Statica*, 38.

47. Arbuthnot, *An Essay*, 82.

48. H. D. Gaubius, "De Regimine Mentis," in *Opuscula Selecta Neerlandicorum de Arte Medica* (Amsterdam: F. van Rossen, 1932), 77:158–60.

49. Galen, *Opera omnia*, 4:807 f.; Charles Victor Daremberg, *Oeuvres Anatomiques Physiologiques et Médicales de Galien* (Paris: Ballière, 1854–56), 1:47–91.

50. Quincy, *Medicina Statica*, vi.

CHAPTER 11

Comments on the German Edition
of Rush's Account of the Yellow Fever

Benjamin Rush's *Account of the Bilious Remitting Yellow Fever as It Appeared in the City of Philadelphia, in the Year 1793* is one of the books—if not *the* book—that accounted for its author's European reputation. In his *Recollections of Dr. Rush* (London, 1815), Lettsom wrote about this book: "When this grand production, uniting in an almost unprecedented degree, sagacity and judgment, first appeared, Europe was astonished" (p. 12). Small wonder, then, that a German translation of the work appeared in 1796, prepared by P. F. Hopfengärtner and J. F. H. Autenrieth.[1] The bulk of the translation was done by Dr. Autenrieth, who also added an appendix entitled "Remarks on the Probable Causes of the Various Forms of the Yellow Fever."[2] Dr. Hopfengärtner wrote an introduction, translated a small part of Rush's text, and added some footnotes to it. The main interest of the book lies in the introduction and the appendix, and the following comments are devoted to these two sections.

Philipp Friedrich Hopfengärtner was born in 1771 in Stuttgart, in the German duchy of Württemberg. Here he became municipal physician in 1794, and in the following year he added the title of court physician. After the death of his wife, he died under tragic circumstances in 1807.[3] Some of his writings dealt with epidemiological subjects, a fact that easily explains his interest in Rush's work on yellow fever. His approach to the latter is rather objective and by no means uncritical. He and Autenrieth decided to translate it in full because they regarded it as a document for the

Originally published in *Victor Robinson Memorial Volume: Essays on Historical Medicine,* edited by Solomon R. Kagan (New York: Froben Press, 1948).

archives of medicine. But on almost all essential opinions expressed with Rush's dogmatic vigor, Hopfengärtner took a skeptical or moderating stand. Rush had considered spoiled coffee and other organic materials the focus of the infection. Hopfengärtner stated that this could not be the main cause of the epidemic and that its importation from the West Indies could not be excluded. If it were objected that yellow fever manifested itself differently in Philadelphia, it might also be said that here the infectious poison found a different climate and different people with a different way of life. Also, Rush's famous treatment of the disease by purging and bleeding did not meet with Hopfengärtner's approval, who considered any standardized form of treatment as ill advised. In particular, he must have been shocked by the radicalism with which Rush had insisted that the people themselves could be entrusted with the cure.

> For a long while air, water and even the light of the sun, were dealt out by physicians to their patients with a sparing hand. They possessed for several centuries the same monopoly of many artificial remedies. But a new order of things is rising in medicine as well as in government. Air, water, and light are taken without the advice of a physician, and bark and laudanum are now prescribed everywhere by nurses, and mistresses of families, with safety and advantage. Human reason cannot be stationary upon these subjects. The time must, and will come, when in addition to the above remedies, the general use of calomel, jalap, and the lancet, shall be considered among the most essential articles of the knowledge, and rights of man.[4]

To these and similar ideas of Rush, the "Hofmedicus" Hopfengärtner remarked: "Consequently, the propagation of medicine among the people as proposed by Rush is obviously a risky undertaking and will, fortunately, remain a pious wish" (p. xxxiv).

Since Hopfengärtner rejected some of the results that Rush had drawn from his theory of the disease, it is not surprising to find that he criticized the theory itself. Or rather, he blamed Rush for the one-sidedness with which he had followed the theory of John Brown (1735–88). The translator himself confessed that the Brunonian system contained several useful indications but declared himself unable to accept it as a whole.

This raises the question as to how Hopfengärtner came to connect Rush with Brown's system, for if we turn to Rush's book we do not find

the name of John Brown in it. "My principal aim," Rush writes, "has been to revive, and apply to it, the principles, and practice of Dr. Sydenham."[5] And indeed, the influence of Sydenham's epidemiological thought is unmistakable throughout. Besides, Rush tried to bolster his famous new practice of excessive purging and bleeding by the authority of this great English physician of the seventeenth century. Thus, we may well ask ourselves how the world came to consider Rush a protagonist of Brown on the basis of this book. And I think that the answer lies in the importance of the concepts of "direct" and "indirect debility"—concepts that Rush owed to Brown.

According to Brown, health and disease depended on the relationship between excitability and stimuli. Excitability was a property of the organism, whereas the stimuli either came from outside the body, like heat, cold, food, contagion, and so forth, or from inside the system, like the passions and emotions. If the degrees of excitability and stimulation balanced each other, the organism was in health. If the stimuli were too weak (e.g., if the body was undernourished), direct debility resulted. But if the excitability of the body was exhausted by excessive stimulation, indirect debility followed. In this case, the body was unable to react normally because it was overloaded by undue stimulation.

The concepts of direct and indirect debility occurred in Rush's works before the book on the yellow fever was written. But in the latter treatise, he gave them a significance that could not be overlooked. He had convinced himself that the contagion produced a debility of the indirect kind, where an excess of stimuli prevailed.[6] Therefore, the patient had to be relieved, as far as possible, of other stimuli. This was done by putting him on a low diet and purging him mercilessly with calomel and bleeding.

It is well known how this "new practice," as it was called, caused a bitter fight among the Philadelphia doctors. Rush became increasingly dogmatic and radical. The calomel pills were sold and dispensed everywhere, people treated themselves, and Rush conceived the idea that in epidemics of plague and yellow fever doctors were altogether superfluous. The organism was always in a state of indirect debility. Therefore, observation of the individual symptoms was unnecessary, and the cure with low diet, purging, and bleeding could be entrusted to the people.

We must keep in mind that the practical inferences drawn from the concept of "indirect debility" belong to Rush, not to Brown. There is little

doubt that Rush spoke the language of Brown,[7] and he used many of his views, including the role he attributed to passions. But there is equally little doubt that Rush was not a mere follower of this system and, before and after the work on yellow fever, stressed his differences from Brown. Thus, in a lecture delivered in 1791, Rush said:

> In the system of Dr. Brown, we find clear and consistent views of the causes of animal life, also just opinions of the action of heat and cold, of stimulating, and what are called sedative medicines, and of the influence of the passions in the production and cure of diseases. But while he has thus shed light upon some parts of medicine, he has thrown a shade upon others. I shall hereafter take notice of all the errors of his system. At present I shall only say, I shall not admit with him, debility to be a disease. It is only its predisposing cause. Disease consists in morbid excitement, and is always of a partial nature: of course I shall reject his doctrine of equality of excitement in the morbid states of the body, and maintain, that the cure of diseases consists simply in restoring the equal and natural diffusion of excitement throughout every part of the system. If Dr. Cullen did harm by directing the attention of physicians, by means of his nosology, only to the names of diseases, how much more mischief has been done by Dr. Brown, by reducing them nearly to one class, and accommodating his prescriptions to the reverse state of the body, of that which constitutes their proximate cause.[8]

And again in 1796 he wrote:

> From the view I have given of the state of the blood-vessels in fever, the reader will perceive the difference between my opinions and Dr. Brown's upon this subject. The Doctor supposes a fever to consist in debility. I do not admit debility to be a disease, but place it wholly in morbid excitement, invited and fixed by previous debility. He makes a fever to consist in a change only of a natural action of the blood-vessels. I maintain that it consists in a preternatural and convulsive action of the blood-vessels. Lastly, Dr. Brown supposes excitement and excitability to be equal in fever. My theory supposes a fever to be the reverse of this. It consists in unequal or divided excitement and excitability. Health consists in the equality and uniformity of them both and the business of medicine, as I shall say hereafter, is to equalize them in the cure of fever; that is, to abstract their excess from the blood-vessels, and to restore them to the other parts of the body.[9]

In between such statements the work on yellow fever appeared, couched in a terminology that pointed unmistakably to Brown at a time when the latter's system stood in the center of medical interest.[10] In Germany, for instance, Brown's *Elements* had been translated by M. A. Weikard in 1795 (i.e., just one year before the translation of Rush's work). The fact that Rush used Brown's ideas as a matter of course and without any outspoken polemic against their author probably served to associate more closely the two men in the minds of the readers. At any rate, Rush's book on yellow fever was considered as supporting Brown's system and was cited as such by Thomas Beddoes in his edition of Brown's *Elements*.[11] And finally it could even happen that Rush's name was deemed memorable in the history of medicine "as the one who in 1793 had introduced Brown's method of treatment in America, induced by the favorable results he had attained with it on the occasion of an epidemic of yellow fever."[12]

Dr. Autenrieth's "Remarks on the Probable Causes of the Various Forms of Yellow Fever" appended to the translation deserves attention on account of its author as well as its contents. Johann Hermann Ferdinand von Autenrieth was born in Stuttgart on October 20, 1772, the son of a civil servant. He studied medicine at the Karlsakademie, where the poet Schiller had studied before him and where his father in 1777 had become professor of political economy. In 1792 he received his medical degree and in 1794 he intended to settle as a practitioner in Stuttgart when his father, embittered by the unfair treatment he had received at the hands of the Duke of Württemberg, decided to emigrate to North America. However, the Autenrieths returned to Württemberg in the following year; the old duke had died in the meantime. Dr. Autenrieth soon became one of the most outstanding physicians and medical teachers of his country. In 1797 he was appointed professor of anatomy, physiology, surgery, and obstetrics at the University of Tübingen. In the following years he taught other branches of medicine and succeeded in reforming the university, whose chancellor he had become in 1822. He still held this position at his death on May 2, 1835.[13]

The appendix embodies some of Dr. Autenrieth's experiences in the United States. We learn from it that he sailed to Baltimore in the summer of 1794 and proceeded further inland without delaying in this city. Lan-

caster, Pennsylvania, seems to have been his residence, though his personal acquaintance extended over New Jersey, Pennsylvania, Delaware, Maryland, and the border of Virginia. In September 1794 he was in Philadelphia, since he tells us that on the twenty-fourth of this month he there examined a yellow fever patient who had come from Baltimore. Shortly afterward, while on some journey, Autenrieth himself fell ill and diagnosed his disease as yellow fever. Far away from any medical assistance, he remembered Rush's treatment and bled himself profusely until he was near fainting. He recovered eventually and returned to Lancaster. "I have reason to believe," he concludes his own lengthy case history, "that, apart from a weakened spleen, my health has not suffered any harm from the severe disease or the equally severe cure" (p. 462 f).[14]

While in Philadelphia, Dr. Autenrieth himself had seen patients recover under Rush's treatment, and he obviously had sufficient faith to try it on himself. It was just the success of Rush's method in North America as contrasted with the experiences of medical men in the West Indies that caused him to raise the question of why the disease, though fundamentally the same, varied in its manifestations and its reactions to treatment. In accordance with Hopfengärtner, he believed that the variations could be explained by differences of land, climate, and inhabitants. From this point of view, he discussed the Delaware Bay and the Chesapeake Bay and their surroundings, particularly the cities of Philadelphia and Baltimore. In Philadelphia he noticed the sudden transition from a cold winter to a hot summer and believed that the climatic influences are further accentuated by the manner of living of a population that has not yet become acclimatized through many generations.

> Very few houses have stoves, but only English fireplaces and the wooden buildings are in any case open to all intruding cold. Even in the most severe winter the inhabitants dress almost as lightly as in the temperate seasons. Nobody, for instance, wears a fur coat. On the other hand, instead of counteracting the disadvantages of the summer heat by a light, mainly vegetarian diet, as the southern Europeans do, they live, the whole year round, chiefly on meat. The high wages and the large number of cultivated acres relative to the population, enable rich and poor alike to live well and, in this respect, diminish the great difference between these two classes that exists in the old world.

All this is further aggravated by so much drinking of tea and—the price of beer being too high—of hard liquor by the Americans, and in the case of workers, by the drinking of the hard, extremely cold pump water even in the greatest heat. The higher classes, moreover, are affected by all the disadvantages of city life (p. 451 f).

All these conditions are supposed to weaken the constitution of the inhabitants of Philadelphia. But Autenrieth considered Philadelphia relatively healthy when compared with Baltimore, among whose inhabitants the transition to the enervation of white persons in very hot countries could be noticed.

[Baltimore] is situated at the end of one arm of the Chesapeake Bay, full of stagnant brackish water not improved here by any running water; and on land, which consists of sand and clay, it is surrounded by swamps. In the back it is hemmed in by wooded hills and towards the southeast only is it open to the exhalations from the bay. Hence it must by necessity be even more uniformly unhealthy than Philadelphia. Swarms of mosquitoes darken the air towards evening. The irritation from their bites with which they cover the foreigners especially, and their buzzing at night completely rob one of any sleep and rest which the oppressive summer heat might possibly allow. One feels one's strength completely exhausted; life seems hardly able any longer to preserve the humors from disintegrating, and the laziness of the inhabitants of hot countries appears very natural (p. 455).

The comparison of Philadelphia and Baltimore and of the diseases prevailing in these two cities serves Autenrieth as a reference by which to gauge what differences climatic and social factors produce in such widely separated districts as Philadelphia and the West Indies regarding the character of yellow fever. His speculations end in contrasting the pathology of the New World with that of the Old. What yellow fever means to the region around the Gulf of Mexico, the plague means to the southern coasts of the Mediterranean. Endemic in these two places, these diseases flare up in epidemics that show striking similarities.

These speculations appeared more convincing to Autenrieth than the opinion of a Dr. Reimarus, whom he cites as believing "that all contagions are beings endowed by nature with a special kind of life and, like plants and animals, divided into definite species" (p. 470 f).[15] To us, how-

ever, his epidemiological speculations are of less interest than the descriptions of land and people of this country and how they appeared to a German physician toward the end of the eighteenth century.

NOTES

1. Benjamin Rush, *Beschreibung des gelben Fiebers, welches im Jahre 1793 in Philadelphia herschte*, Aus dem englischen übersezt und mit einigen Zusäzen begleitet von P. Fr. Hopfengärtner and J. F. H. Autenrieth (Tübingen, 1796).

2. "Bemerkungen über die wahrscheinlichen Ursachen der verschiedenen Formen des gelben Fiebers von Dr. Autenrieth," in ibid., 439–72.

3. E. Gurlt and A. Hirsch, *Biographisches Lexikon der hervorragenden Aerzte aller Zeiten und Völker* (Wien, 1886), 3:273.

4. Benjamin Rush, *An Account of the Bilious Remitting Yellow Fever, as It Appeared in the City of Philadelphia, in the Year 1793* (Philadelphia, 1794), 327 f.

5. Ibid., 337.

6. Ibid., 224–27.

7. N. G. Goodman, *Benjamin Rush* (Philadelphia: University of Pennsylvania Press, 1934), 231.

8. Benjamin Rush, "On the necessary connexion between observation, and reasoning in medicine," in *Sixteen Introductory Lectures* (Philadelphia, 1811), 11 f.

9. Benjamin Rush, "An inquiry into the proximate cause of fever," in *Medical Inquiries and Observations* (Philadelphia, 1796), 4:147 f.

10. Cf. B. Hirschel, *Geschichte des Brown'schen Systems und der Erregungstheorie*, 2nd ed. (Leipzig, 1850).

11. John Brown, *The Elements of Medicine*, a new edition, revised and corrected with a biographical preface by Thomas Beddoes (Portsmouth, N.H., 1803), lxxvii, lxxxii.

12. Gurlt and Hirsch, *Biographisches Lexikon*, 5:126.

13. The biographical sketch is based on the articles on Autenrieth in *Allgemeine Deutsche Biographie* (Leipzig, 1875), 1:696–96.

14. Besides bleeding himself, Autenrieth had taken rhubarb, tartar emetic, and "red china bark." Dr. Clifford B. Farr, in a personal communication, suggested to me that Autenrieth's case history pointed to malaria rather than yellow fever.

15. Autenrieth here refers to J. A. H. Reimarus (1729–1814), a physician and scientist of Hamburg, who expounded these ideas in his preface to the German edition of d'Antrechau's *Merkwürdige Nachrichten von der Pest in Toulon* (Hamburg, 1794); cf. G. Sticker, *Abhandlungen aus der Seuchengeschichte und Seuchenlehre*, vol. 1: *Die Pest*, 2nd Theil (Giessen, 1910), 16 f.

A Miscellany

CHAPTER 12

History and Prophecy
Meditations in a Medical Library

The opening of a new medical library is an important event in the life of medicine. A library is not only a gathering point for students, professors, practicing physicians, and others interested in medicine; it is also the point where past, present, and future meet. The library makes available knowledge accumulated up to the present day, knowledge that we assimilate with an eye to things to be done in practice or in research, be it today, tomorrow, or at some distant time.

Thus, a medical library is a monument to the Hippocratic aphorism "Life is short, the art is long."[1] It is a monument to the short lives of those whose deeds and thoughts have passed into the stream of the medical art that flows on endlessly. If their names are forgotten, their minds are still present. Every textbook of physiology incorporates the ideas of William Harvey, even if the name of the discoverer of the circulation of the blood is not mentioned. We all try to leave our mark, to have something of ourselves pass into the future. We hate to think that we may have labored in vain and that the future readers in the library will find no trace of us, not even an entry in a catalogue.

Because we all have a share in the preservation of human thought, we recoil from the destruction of libraries and of books. Some thirteen hundred years ago, when the Arabs conquered Alexandria, the victorious general wrote the caliph, the religious head of Islam, asking what to do with the famous library. The caliph replied:

Originally published in *Bulletin of the History of Medicine* 49, no. 3 (fall 1975). © The Johns Hopkins University Press.

> Regarding the books you mentioned: If their content is consistent with the book of God, then the book of God suffices. But if their content is contrary to the book of God, there is no need of them. Therefore have them destroyed. Thus, Amr ibn Al'ās began to distribute them among the baths of Alexandria, and to have them burnt in their fireplaces. The number of baths at that time was pointed out to me, but something has made me forget it. Then it is stated that the burning lasted over a period of six months. Hear what happened and be astonished![2]

The story is a legend, the library had been burnt before, and so thoroughly that few books could have remained. But the fact that such a legend arose at all and that scholars have taken great pains to disprove it shows how sensitive we are to the destruction of the documents of human aspirations and thoughts. The reason for our aversion was, I think, well expressed by John Milton, when he wrote "who kills a man kills a reasonable creature, Gods Image; but hee who destroyes a good Booke, kills reason it selfe, kills the Image of God as it were in the eye."[3]

A medical library, then, spans past and future. It tells us not only what medical men thought was true and good for their times; it also allows us glimpses into what they thought the future held for them and for those after them. We are in a position to judge their predictions, to see whether they have come true or not, and to meditate on their prophecies.

Prediction is part of the physician's trade, only he calls it *prognosis.* The physician was sometimes associated with the soothsayer, with the difference that the physician's predictions were expected to be based on experience, those of the soothsayer on portents and on inspiration. But clinical signs were not always distinguished from magic portents or divine inspiration. One of the oldest of the ancient Greek poets invoked the Muses "with their songs, telling of things that are and shall be and that were aforetime."[4] Likewise, one of the Hippocratic books, written about 400 B.C., opens with the words:

> I hold that it is an excellent thing for a physician to practice forecasting. For if he discover and declare unaided by the side of his patients the present, the past, and the future, and fill in the gaps in the account given by the sick, he will be the more believed to understand the cases, so that men will confidently entrust themselves to him for treatment.

This is the beginning of the book that contains the famous description of the face of a man about to die, the *facies Hippocratica:* "Nose sharp, eyes hollow, temples sunken, ears cold and contracted with their lobes turned outwards, the skin about the face hard and parched, the colour of the face as a whole being yellow or blackish."[5] You will notice that for Hippocrates prognosis was directed not only to the future but to the past and present as well. Ancient physicians were not identified by a diploma or a license; they had to gain the confidence of their clientele by being able to tell what was wrong and what had been wrong with the patient, data that could be verified. Something of this still lives on in us, if only in the joke of the man whom the dentist asked which tooth hurt and who replied, "You are the doctor, so you ought to know!"

Of course, every physician knows that prognosis is full of uncertainties. Even the most experienced physician can err. In the same aphorism that I mentioned before, Hippocrates warned, "experience is treacherous."[6] Prognosis may be safer today, when there is less reliance on purely clinical experience and more support from experimental science, with its abundant supply of laboratory data. But let me say that the Greek word in the Hippocratic aphorism that I just quoted as meaning treacherous can also be translated as "perilous" and experience as "experimentation," so that the aphorism takes on a different meaning: experimentation is perilous. Thereby the ancient commentators meant that trying out unknown remedies may be the patient's undoing.[7]

Indeed, we are inclined to attribute shortcomings in our ability to predict the future to imperfect knowledge, and this holds true with regard to the future of not only the individual patient but medicine as a whole. We do not yet know how to cure multiple sclerosis, as yet we have not conquered inoperable cancer, and so on.

When we say that we do not yet know something, we express our confidence that some day we shall know it. We take it for granted that it can be known and that we have not yet progressed far enough on the road to discovery. This is an attitude shared by physicians and laymen. The layman indignantly asks his doctor, "Why can you not cure this disease?" To which, as you all know, the proper answer is, "They have not yet found a remedy for it." It is reflected in the enthusiastic financial support given for an all-out fight against this or that disease. We are confident that a way

can be found to eradicate it and that, given the means, that way will be found.

If due allowance is made for exaggerated zeal, which can lead to disappointments, we have every right to expect of scientific progress that it will lead to an increase in what can be done in medicine. The achievements of the last forty years confirm this clearly. To read the bare catalogue of all the discoveries and inventions that have been made since the introduction of the sulfa drugs would alone take up the rest of the time at our disposal. If you will allow me to delve once more into my memories and the days of Latin tags, we used to define internal medicine as *circumstare et verba facere*, "to stand around and make words." Even then, in the mid-twenties, that was not quite true: insulin had been discovered and liver diet in the treatment of pernicious anemia was proving helpful. But compared to the pace with which discoveries of new remedies now follow one upon the other, it was a time when progress was hoped for but not yet counted on. I do not remember having heard any of my teachers tell us that prolonging the life of a hopelessly ill patient gave him the chance of living long enough to benefit by the discovery of an effective cure.

Nearly a hundred years ago a surgeon, a pioneer in the introduction of antiseptic surgery, remarked that the surgeon no longer was allowed to hope for a successful outcome of an operation; rather "he was the manufacturer from whom a good product was *demanded*."[8] Such a statement would not surprise us if made today. But having been made in 1881, it sounds more like a prophecy come true than a sober description of the prevailing state of affairs. Indeed, I think that a prophetic note was struck when the surgeon was compared to a manufacturer and, by implication, the public to consumers. The statement anticipated a future not only of great reliability but also of commercialized medical care.

I said that our confidence in predicting the outcome of individual disease and in the progress in the prevention and treatment of diseases rests largely on the greater precision that science has given to medicine. But our last example suggests that we must not think of natural science alone; we must include the social sciences as well. Highly personal as the relationship between doctor and patient is, it has its place within a framework of ever changing social and economic conditions. This is true not only of medical care administration in this country but also of the worldwide effects of medicine. We are appalled at some of the consequences of mod-

ern medicine and public health for the population of underdeveloped countries. Mortality rates have dropped, but misery rates have not. Famine, malnutrition, and disruption of traditional ways of life tend to counteract the benefits medical science has brought. To some extent this might have been foreseen, and perhaps it actually did occur to some people. At any rate, it is a warning that medical science and social science have to go together and that the predictive ability of both is needed.

Thus far, our discussion has envisaged the kind of predictions that we make and must make. We cannot live without predicting, without planning for tomorrow and relying on our plans' materializing. This is a truism for life in general. We are confident that we shall go to work, that we shall meet people with whom we have appointments; we resist the temptation to buy a packet of cigarettes because of the chances of developing cancer of the lung; we plan an operation and make sure that an operating room will be available; we plan an experiment and consider the consequences of positive and negative results; we participate in a conference on the medical curriculum—probably not for the first time—and wonder what the changes will mean for the next generation of physicians. We could go on and heap example on example, showing that we could not live without the firm belief that tomorrow the sun will rise and bring us another day.

We may overrate the knowledge we possess. Long ago, William Gowers, the famous British neurologist, remarked that "our knowledge is enough to obscure our ignorance."[9] He was speaking about fainting, and you will remember that Gowers's syndrome is an eponym for the vasovagal attack. But I am afraid that Gowers's remark holds good on a much broader level. Until about ten years ago, it was fashionable to talk about preparing the world for the year, let us say, 1980 or even 2000. We thought we owed this to the future. We are no longer quite so confident that we have a clear picture of what will be needed. To give but one important example: the rate of increase in population has changed. The population explosion continues in some countries, whereas in others, particularly those of the industrialized West, it has abated. The population of the world or even of one country is a basic element in calculating what is needed and what can be done. We know some factors that have countered the steep increase in some countries: successful planning, simple and more reliable contraceptives, and, more recently, less favorable economic conditions. But I doubt that we are really capable of estimating the rise and

fall of procreative desire (i.e., the wish to have many or few children). Moral and religious motives play a great role, and who will be bold enough to calculate the moral and religious feelings of future man?

Bold enough or not, in this kind of prediction we no longer rely on knowledge but on a vision of things to come. We have a very ambivalent attitude toward vision. On the one hand, we praise "a man of vision," and we blame a person for having a lack of vision. This is natural because some degree of vision is needed for all we undertake. The person who buys a bottle of whiskey has a vision of the party he is going to give, and if a person of great vision, he also buys a bottle of aspirin for the ensuing hangover. On the other hand, we speak of a visionary in a derogatory way; visions dissociated from reality are abnormal—they belong to true prophets or to the mentally ill. We easily call a man a crackpot if his visions seem fantastic to us. Perhaps we ought to keep in mind what some Englishman said about crackpots long ago, "They are certainly cracked, but the cracks let in light."[10]

However, I am not thinking here of either extreme: the trivial or the abnormally exalted. I am thinking of visions that guide man in science and in art. Three hundred fifty years ago, the philosopher and mathematician Descartes had the vision of a mechanically determined world in which even man's body was nothing but a machine, the soul alone being capable of sensation and of will. This vision is still alive today: in spite of psychosomatic medicine, we have great difficulties in overcoming the dichotomy of body and mind. Paul Ehrlich, the founder of modern chemotherapy, had a vision when he searched for specific, chemically prepared drugs that would act like magic bullets against pathogenic micro-organisms.

From these examples of the tremendous power of vision, we now turn to a scrutiny of its role. Permit me to read to you the following passage from a medical address, without naming the speaker:

> The advances in our knowledge in medical science within the last forty years are without a parallel in any age. Never was the medical profession so busy and industrious, so zealous and enthusiastic, so honest and exact in its views and its results, as it is at the present moment. It would also seem as if the millennium were actually close at hand. Look where we may, progress, rapid and brilliant, nay, absolutely bewildering, literally stares us in the face, and challenges our respect and admiration. One is almost ready to exclaim. "Behold, all things are new!"[11]

These lines sound like the praise so often given to the progress of medicine during the last forty years, since the introduction of the sulfa drugs. Well, the speaker was Samuel David Gross, and the words are from the introduction to a course of lectures he gave at Jefferson Medical College in 1867!

What Gross here presented was also a vision, not of the future, but of his own time. How was it possible to be so enthusiastic about 1867, when our medical educational system was at a very low level, when surgery was hamstrung by the constant threat of infection, when even aspirin did not yet exist, leave alone X-rays, insulin, sulfa drugs, the antibiotics, and the use of hormones and vitamins!

Gross—he lived from 1805 to 1884—was probably the greatest American surgeon of his time. He had witnessed the rise of pathological anatomy in Paris, London, Dublin, Vienna, and Berlin and had himself written one of the best textbooks on the subject. Inhalation anesthesia had become available some twenty years before his address. Cellular pathology and experimental medicine were novelties. Put together, these and other items make us understand why Gross thought more highly of medicine in 1867 than we do. But it is, I think, fair to say that his vision of the contemporary status of medicine blinded him to its faults and weakened his perception of the future. Though he himself stated that "the materia medica has been purged of many of its crudities, and attained a more exalted position," he yet pointed to "a large octavo volume of upwards of seven hundred pages upon what are called new remedies," which a colleague of his had written.[12]

Whatever these new remedies were, it is safe to say that most of them have gone with the wind; not because they were all tried out and found wanting, but because modern pharmacology swept away innumerable drugs for which no other evidence than tradition could be cited. In 1867, Gross could not know the results of modern pharmacological science. But he lived in the days of strong therapeutic skepticism and even nihilism. Men like Oliver Wendell Holmes had come back from Paris with a deep-seated distrust of traditional therapy. Their teacher, Pierre Louis, had preached the gospel of the numerical method (i.e., of the statistically evaluated therapeutic trial). You have probably heard Holmes's often quoted remark "that if the whole materia medica could be sunk to the bottom of the sea, it would be all the better for mankind,—and all the worse for

the fishes."[13] In Vienna, therapeutic nihilists had agitated for suspending all therapy, letting the patient have good nursing care, and concentrating on scientific research that would create a reliable pharmacology. In 1872 the first volume of the *German Archive for Experimental Pathology and Pharmacology* appeared, the organ of the new experimental science of pharmacology.

Gross misread the signs of the time when, in his essay, he claimed that the statistical method was defunct, even though its founder, Pierre Louis, was still alive.[14] And in 1876, nine years after Lister had introduced antiseptic surgery, Gross wrote:

> Little, if any faith, is placed by any enlightened or experienced surgeon on this side of the Atlantic in the so-called carbolic acid treatment of Professor Lister, apart from the care which is taken in applying the dressing, or, what is the same thing, in cleaning away clots and excluding air from the wound; an object as readily attained by the "earth dressing" of Dr. Addinell Hewson of Philadelphia, and by oil dressing . . . which I have myself employed for many years, with signal benefit, in nearly all cases of wounds under my charge, whether the result of accident or design.[15]

Clearly, Gross had failed to understand the principle underlying Lister's use of carbolic acid. Was Gross then a short-sighted man? Not necessarily. His misjudgments show how easy it is to be right with the wisdom of hindsight and how difficult when you are in the midst of new developments. When Gross took a dim view of the statistical method, he was in excellent company. Only two years before, Claude Bernard, in his *Introduction to Experimental Medicine,* had led a sharp attack on statistics as a source of medical knowledge. When Gross disparaged antiseptic surgery, he was probably guided by his own experiences. Surgeons who did not operate in crowded hospitals, where infection was rampant, who did not hurry from pathological autopsy to the operating room or to the delivery room, who were fastidious in washing their hands before an operation and careful in the attention they gave to their patients had results that could compete with those of Lister. We praise Lister, who had applied Pasteur's germ theory to surgical infection, and we praise Semmelweiss, who in 1847, on the basis of statistics, proved that puerperal fever was an infection transmitted by the hands and instruments of the physicians and who advised washing the hands in calcium chloride solution. The results

were there for all to see, but before Pasteur the scientific reasons for such successes were not clear. Thus, in Vienna, the slogan was coined: "Wait and wash!"[16] As to the germ theory itself: Robert Koch, the founder of modern bacteriology, was to show that no disinfectants would rid the hands completely of bacteria.

The real history of medicine is much more complicated than we historians usually portray it in our textbooks. The vision of a genius needs a certain single-mindedness to cut through the bewildering, contradictory evidence of contemporary life; it needs a compulsion to follow the vision and thus comes perilously close to the monomania of the crackpot. Semmelweis called the obstetricians who disregarded his work murderers. To them in turn his zeal must have appeared that of a fanatic.

Gross, I think, illustrates the danger of being dazzled by too bright a vision of one's own time, for it breeds conceit and blinds one to possible crosscurrents. Imperceptibly, the proud medical school of Paris began to be overshadowed by Germany, with its strong scientific institutions. German medicine, in turn, was still basking in its own glory when leadership was slipping away to the United States. History becomes dangerous when misused for self-congratulation.

But let us turn from the vision of one's own time to that of the future. In 1913, William Osler gave an address to the Yale students. This address, *A Way of Life*, is among his best known. He urged the young people to live "a quiet life in day-tight compartments," that is, to do the tasks that were before them, rarely looking back to take stock and not indulging in daydreams about the future. Life, so Osler thought, was "a straight, plain business," and he had a vision of the future of the students he addressed: "In my mind's eye, I can see you twenty years hence—resolute-eyed, broad-headed, smooth-faced men who are in the world to make a success of life."[17]

But in 1914, World War I broke out. It claimed the life of Osler's only son and, I am afraid, of many a Yale man who never lived to fulfill Osler's prophecy. To most people of that time, that war came as a great surprise, for the world seemed stable, progressing toward an ever better material and social life. A war as barbarous as World War I proved to be was almost unthinkable. We shall probably not be guilty of the same oversight as Osler, for we are accustomed to violence of all kinds, and we have long disabused ourselves of the idea that the world is bound to become more

civilized. Does that mean that we are better prepared to have visions of the future that are not vitiated by the unforeseen? I doubt it because, guard as we may against the unforeseen, we cannot eliminate the unforeseeable.

In theory we all acknowledge the existence of the unforeseeable; it sounds trivial to insist on it. But in practice we disregard it again and again; rarely do we add an old-fashioned *deo volente,* if God will, to our plans. I have already mentioned the past fashion of preparing the world for what was to come in the year 2000.

Let us take a few examples of the unforeseeable. In 1491, nobody could foresee that in the next year a new continent would be discovered and that this discovery, together with that of the sea route to India, would create an economic revolution and a re-orientation of trade routes and commerce. All predictions made in 1491 for the state of the world in 1550 were bound to be wrong.

The invention of such instruments as the telescope and microscope around 1600 was another invasion of the unforeseeable—new instruments that made new regions accessible to our senses. With the aid of the telescope, Galileo showed the moons of Jupiter. Looking through this man-made tube with its lenses disproved finely spun theories. The same was true when the microscope showed organisms and organic structures that could not have been anticipated.

Just as we have become prepared for political upheaval, so also we have become accustomed to seeing new discoveries upset our medical theories. But that does not yet mean a ready acknowledgment of the unforeseeable. In the 1970 edition of a leading textbook of pharmacology, we read, in the chapter on estrogens, of "a trial of diethylstilbestrol in doses as large as 2 g daily" during pregnancy. The author adds, "Although this form of treatment did not prove effective in toxemia, its harmlessness was reassuring."[18] Perhaps he was thinking only of late pregnancy; at any rate, in early 1971 the first report came in of vaginal cancer in girls between fourteen and twenty-two years of age, most of whose mothers had been treated with this drug in early pregnancy.[19] Who could have predicted this? Neither animal experiment, nor the usual clinical trial, nor an early epidemiological survey of a drug will indicate effects that will reveal themselves decades later. Only what I like to call "the historical trial" will eventually record what drugs do and do not do.

I call such trials *historical* because the word implies that they are made a

considerable, yet undetermined, time after the drug has come into use. And as history is constantly rewritten, it is in the nature of the historical trial that its results may also vary at different times. Its only drawback is that the patients, too, may have departed into history long before the trial takes place.

Visions need not be either true or false. There is perhaps no better example of the truth *and* fallacy of vision than that offered by Daniel Drake in his *Principal Diseases of the Interior Valley of North America.* This is one of the very great works on medical geography; it describes what we now call the Middle West, its geographic and climatic features, its inhabitants and their cultural characteristics, and, as the title says, their diseases. After seven hundred pages of descriptive text, Drake pauses. "All peculiarities of constitution, both corporeal and mental," he thinks, "exert a modifying influence on disease." The peculiarities are as yet not fully developed, but we may infer the effects that will appear in full development to "our distant successors." And so Drake ventures upon a grandiose assessment of what he sees going on and presents a vision of what will come to pass: "A synthesis of varieties and races is going on; and the result, I may here repeat, must be a new national constitution—physical and mental—of which the Anglo-Saxon, itself a compound, will be the basis and the governing element."[20]

Until about twenty years ago this could be called a prophecy come true. It was a superb anticipation of America as "the melting pot" of peoples of different nationality, molded into a nation with its own physiognomy and led by the Anglo-Saxon element. Following the example of my teacher, Henry E. Sigerist, I presented this passage to my students as one of the great visions in American medicine, seen by a man who embodied the pioneer spirit of his time, who had real insight into its historical meaning. Were we right?

Dr. Sigerist died in 1957; until then, Drake's vision could still be considered a true prophecy of what his "distant successors" would see. But we here in 1975 are even more distant successors, and to us Drake's vision no longer seems indisputable. The slogan of "the melting pot" is less frequently used now, when much weight is placed on ethnicity and the Anglo-Saxon element considered one among others.

Alerted by this consideration, let us continue with Drake's words. He admits that the diseases of the future "under the joint influence of min-

gled blood, of climate, water, occupations, modes of living, customs, and moral, social, and political influences, cannot be specified." Again, let us admire the sweep of his mind, the wide range of factors he took into account 125 years ago, long before sociology had become fashionable and "ecology" a slogan. I wish he had stopped on this negative note. But Drake was a courageous man who, in his turbulent life, had fought many battles. And so he hazarded a few predictions, few of which have come true to date. He foresaw a decrease in "autumnal fevers," and, as far as this meant malaria, he was right. But he thought that typhus and typhoid fevers would become more prevalent—which they have not. "Consumption and scrofula will increase"; perhaps they did for a short time, but they began to decrease before the end of the nineteenth century, and scrofula (i.e., tuberculous lymphadenitis) is, I believe, rarely seen. The last of the seven predictions is of special interest: "Mental alienation will be more frequent." Judging by the number of patients in mental hospitals and under psychiatric care, this is certainly true. But to a large extent this is a consequence of much greater attention given to even minor mental deviations. One of the main scourges of the nineteenth century, general paralysis, a late form of cerebral syphilis, has diminished dramatically.

As the example of Drake shows, prophecies have a life of their own. Unless cast for a particular year or a particular day, their validity depends very much on when and by whom they are judged. Historical judgments vary; this is true not only of Drake's vision, but of others as well. The Reverend Cotton Mather, notorious for his role in the Salem witchcraft trials, wrote a book that he called *The Angel of Bethesda,* a peculiar mixture of Puritan theology and medical ideas very new in his time. Regarding the latter, he dealt at length with the animalcular theory of disease. Oliver Wendell Holmes, who had read the manuscript, took a dim view of the book and remarked, "It is well that the treatise was never printed." He acknowledged Mather's role in the introduction of variolation, the forerunner of smallpox vaccination.[21] But the animalcular theory of disease did not impress him.

In 1869, the year of Holmes's lecture, animalcular theories of disease were still considered fantastic by many of those who could not follow Pasteur. Now, however, Cotton Mather is praised as a man of remarkable vision, and *The Angel of Bethesda* has found its way into print.[22] Indeed, "books have their fate according to the comprehension of the reader."

With this final quotation from an ancient grammarian,[23] we may end our meditations, our rambling thoughts evoked by old books and old sayings. A medical library, and especially a rare book room, is a good place to meditate. The busier our daily life, the more intent we are on gathering information, the greater the need, I think, for occasional consultation of medical men of the past. If we approach them not as representatives of imperfect knowledge and of a dead science long overcome but as human beings who were as alive as we are and who hoped for the future as we do, they will speak to us. And what do they tell us?

Allow me to give a personal answer. People who hear that I am a historian of medicine sometimes say that they prefer to look forward, to the future. But, I think, the historian also looks to the future; he wants to know what predictive power we human beings possess, and he thinks that past experiences are enlightening. They can help us not to venture forth foolhardily nor to lose courage if our expectations fail to be realized—in short, to develop a feeling for the fate of human affairs. As Milton said, "Many a man lives a burden to the earth; but a good Booke is the pretious life-blood of a master spirit, imbalm'd and treasur'd up on purpose to a life beyond life."[24]

NOTES

This chapter was originally presented as an address, May 15, 1975, on the occasion of the dedication ceremonies of the Raymond H. Mulford Library of the Medical College of Ohio at Toledo.

1. *Aphorisms* 1.1; *Hippocrates,* trans. W. H. S. Jones (Loeb Classical Library), 4:99.

2. Ibn al-Qifti, *Ta'rih al-hukama',* ed. Julius Lippert (Leipzig: Dieterich, 1903), 355.20–356.2.

3. John Milton, *Areopagitica,* ed. Hales (Oxford: Clarendon Press, 1904), 6.

4. Hesiod, *Theogony* 37–38; *Hesiod,* trans. Hugh G. Evelyn-White (Loeb Classical Library), 81.

5. *Prognostic* 1 and 3; *Hippocrates,* trans. Jones, 2:7, 2:9 (slightly changed).

6. *Aphorisms* 1.1; ibid., 4:99.

7. See Galen, Commentary on the aphorisms of Hippocrates 1, in *Opera omnia,* ed. C. G. Kühn (Leipzig: Cnobloch, 1821–33), 17B:353–54.

8. R. Volkmann, *Die moderne Chirurgie* [paper read in 1881], *Sammlungen klinischer Vorträge* 221 (Chirurgie no. 70): 2.

9. William R. Gowers, *The Borderland of Epilepsy* (Philadelphia: Blakiston, 1907), 1.

10. J. Moreau (de Tours), *La psychologie morbide dans ses rapports avec la philosophie de l'histoire* (Paris, 1859), 205.

11. Samuel David Gross, *Then and Now: A Discourse Introductory to the Forty-third Course of Lectures in the Jefferson Medical College of Philadelphia* (Philadelphia, 1867), 4.

12. Ibid., 10.

13. Oliver Wendell Holmes, "Currents and counter-currents in medical science," in *Medical Essays, 1842–1882* (Boston: Houghton Mifflin, 1891), 203.

14. Gross, *Then and Now*, 34.

15. Samuel David Gross, "Surgery," in *A Century of American Medicine, 1776–1876*, ed. Edward H. Clarke et al. (Philadelphia: Henry C. Lea, 1876), 213.

16. Erna Lesky, *Ignaz Philipp Semmelweiss und die Wiener medizinische Schule* (Vienna: Böhlaus Nachfolger [Kommissionsverlag der Oesterreichischen Akademie der Wissenschaften], 1964), 51.

17. William Osler, "A way of life," in *A Way of Life, and Other Selected Writings of Sir William Osler* (reprint New York: Dover, 1958), 243, 249.

18. Louis S. Goodman and Alfred Gilman, *The Pharmacological Basis of Therapeutics*, 4th ed. (New York: Macmillan, 1970), 1547.

19. Arthur L. Herbst et al., "Adenocarcinoma of the vagina," *New England Journal of Medicine* 284 (April 22, 1971): 878–81.

20. Daniel Drake, *A Systematic Treatise, Historical, Etiological, and Practical, on the Principal Diseases of the Interior Valley of North America* (Cincinnati: Winthrop B. Smith, 1850), 1:701.

21. Holmes, "The medical profession in Massachusetts," in *Medical Essays*, 360, 362 f.

22. See Otho T. Beall Jr. and Richard H. Shryock, *Cotton Mather: First Significant Figure in American Medicine* (Baltimore: Johns Hopkins Press, 1954), esp. 87–126; Cotton Mather, *The Angel of Bethesda*, ed. Gordon W. Jones (Barre, Mass.: American Antiquarian Society and Barre Publishers, 1972).

23. Terentianus Maurus, *De syllabis*, in *Grammatici latini*, ed. Heinrich Keil (Leipzig: Teubner, 1874), 6:363: "Pro captu lectoris habent sua fata libeli."

24. Milton, *Areopagitica*, 6.

CHAPTER 13

On the Reading of Medical Classics

In his *Areopagitica* John Milton wrote, "For Books are not absolutely dead things, but do contain a potencie of life in them to be as active as that soule was whose progenie they are; nay they do preserve as in a violl the purest efficacie and extraction of that living intellect that bred them."[1] To modern medical library practice, these well-known words constitute something of an embarrassment. The huge increase in the production of books and the speed with which contents become obsolete have forced libraries to handle their books as subject to aging and even to death. Active life is limited to about twenty years, when they are "cut off" and retired as elderly to less accessible areas. Those that pass through the process of aging are resuscitated in the historical collection.

In past times, scientific progress was slow compared to ours; important new discoveries were made at lengthy intervals, and the conservatism of many universities prolonged the lives of authoritative works. In 1644, when the *Areopagitica* was published, not only Hippocrates and Galen but many other ancient authors and even Avicenna were still printed and studied as sources of medical knowledge and science. The final break with the ancient past, its relegation to history, belongs to the mid-nineteenth century. Even then the content of medical books could count on a lengthy period of acceptance. For instance, in 1866, Austin Flint's *A Treatise on the Principles and Practice of Medicine* was still oriented toward the Paris school and showed little regard for contemporary German achievements which, among other things, had produced cellular pathology.

It was formerly, therefore, relatively safe to entrust new scientific dis-

Originally published in *Transactions and Studies of the College of Physicians of Philadelphia*, 4th series, 42, no. 1 (July 1974).

coveries to books, as did Vesalius, Harvey, Auenbrugger, and Jenner. Not only was the danger of concomitant discovery by others smaller; a book could also count on readers with sufficient leisure to peruse it. The feeling of urgency to acquaint the world with one's discoveries and to learn what others had discovered as quickly as possible may not have been as strong as it is today; the less developed means of communication may not have fostered the fear of missing the latest contribution; the way of life of a less industrialized society may not yet have bred the feeling that time was of the essence and the concomitant unrest that overtakes us if we are expected to follow arguments in discursive books not packed with information. Whatever the causes may have been—and they are hard to assess—the compulsion to obtain the most recent information did not manifest itself as strongly as it does today.

By Jenner's time, medical journals were already well established and had replaced the epistolary commerce by which scholars of the seventeenth century let their colleagues and friends know of their novel thoughts and discoveries. The frequently appearing issues of a journal brought information more quickly than a book, but journals, too, are now being overtaken by the torrent of scientific events, for rarely can a contributor count on having his article published within less than a year's time. Journals become volumes that also crowd the shelves, and their contents, too, become obsolete, though the continuity of a journal, and hence the frequent need to go back to an older article, delays the process of aging.

Nor has the progress of technological science that has created some of the difficulties for our libraries bypassed the libraries themselves. From places where books and journals are deposited and catalogued, libraries are turning into information centers, places where ingenious mechanical processes retrieve information, not only saving the reader's time but also making the coverage more extensive. Audiovisual devices offer information on selected topics in concentrated form; they also relieve the viewer of the need to read. There is a constant endeavor to save space by the use of micro-cards and microfilms, and the ideal goal would be reached if the volumes of books and journals could altogether be replaced by less space-consuming material. Indeed, some types of books are already facing doom.[2]

This discourse on the problems of our libraries, by one so little competent for it, is intended to illustrate the intellectual situation in which we

find ourselves. We need ever more information; we have to make room for it in our brains, no less than in our libraries, and we also have to rid ourselves of what is no longer valid and had best be forgotten, lest it do harm. Yet we realize that we cannot destroy all the informative books and journals without obliterating the traces of our civilization and of our own medical activities. In course of time, all books (and for simplicity's sake the term is here to include journals as well) assume some degree of historical value. From giving medical information to us, they change to giving information about us. Thus they reach the heaven of the historical collection. Not all elderly books are saved, however; many of them leave the purgatory of the distant shelves only to disappear forever in such hell as fire, pulping, or other tortures provide.

The great need for increased information, without which our economic, political, and administrative life could not function and without which science and medicine would stagnate, tends to encourage a cult of information that overlooks the fact that even knowledge needs more than information. It needs an integration of data into a comprehensible context, and what makes things comprehensible depends very much on the assumptions and rules we acknowledge when information reaches us. Information must be expressed and it must be organized, both of which can be done well or badly. Even the most purely informative book containing nothing but statistical material can be good or bad, depending on the way it handles that material.[3] Moreover, the material may be correct but irrelevant (i.e., either not pertinent to the theme of the book or related to matters to which we are not willing to assign much value). Or the book is worthless because the material it contains, even if correct, is a mere rehash of earlier publications.

Obvious as these remarks may be, they remind us that every book contains something that is not just informative and that interferes with the prearranged march from young and "active" to "elderly" and then to "old." Some books are judged so worthless that they are not even admitted to the home of the elderly. (Paradoxically, they may then become valuable as rarities.) On the other extreme stand those books that are deemed "timeless," the classics with which we are now concerned.

What is essential for making a medical book, or article, a classic? Examples from nonmedical literature are helpful in the search for an answer. For religious persons, the Bible is "the book" because it reveals divine

truth, and even others will admit that it has much wisdom to impart. Wisdom is more than knowledge, or all who know much would also be wise, which unfortunately is far from being the case. Few will be insensitive to the moving power of prophetic speech, to the feeling of human grandeur and vulnerability conveyed by a Greek tragedy, or to the beauty of Shakespearian sonnets. Few will escape meeting in real life such fictitious characters as Don Quixote, Hamlet, Pickwick, or Babbitt.

But long ago, Claude Bernard pointed out the difference between personal contributions in the arts and the sciences. Sciences being cumulative, a discovery if confirmed is assimilated into later work, without leaving traces of its origin. Not so in the arts, where Plato's philosophy, Shakespeare's plays, and Dickens's characters do not become dissolved in the works of later authors. As Claude Bernard put it: "The past keeps all its worth in the creations of art and letters; each individuality remains changeless in time and cannot be mistaken for another."[4]

Medical classics probably would not exist if they were to be read for their substantive content only. I doubt that a modern neurologist will learn much about epilepsy from Hippocrates's *On the Sacred Disease,* that Fracastoro's *Syphilis sive De morbo gallico* has much to teach about the clinical manifestations of syphilis as seen today, or that Virchow's *Cellular Pathology* contains much that a modern textbook of pathology has missed. To be sure, *On the Sacred Disease* is the first monograph on epilepsy and *Cellular Pathology* the first work basing human pathology on the life of the cell. To be "the first" may make a book very interesting from a historical point of view, and I am the last to discourage historical interest. Nevertheless, a historically important book is not necessarily a classic.

Because medical classics, I believe, are not read primarily for the sake of their substantive content, they are not widely read altogether outside of required class work, with which I am not concerned here. How much they are read voluntarily is hard to gauge. The multitude of reprints (facsimiles, paperbacks, anthologies) points in one direction; my personal impression, I regret to say, points in another. However that may be, it will hardly be claimed that medical men are given to avid reading of medical classics. If this evaluation of the state of affairs is correct, there should be no hesitation on the part of many physicians to proclaim the reading of medical classics a waste of precious time. Yet I do not believe that such a proclamation would find a large number of subscribers. Unless the hes-

itation is attributed to fear of being held lacking in cultural polish, we must look for a motive strong enough to prevent disrespect. In spite of the absence of perceptible immediate usefulness, there must be a belief, or at least a suspicion, that reading such books or articles carries its reward in itself and is, perhaps, indirectly useful.

Since this is not written to promote the classics among the heathen, telling them why they have to read them, the analysis had best begin with an incontrovertible fact: There are those who derive enjoyment from such reading, and there are others who do not. The number of persons incapable of finding a single book that they would be willing to accept as a classic is probably very small. But education and reading habits may stand in the way of accepting many an established classic. There is nothing here to single out medical works from others. The *Oedipus Rex* of Sophocles is an established classic and may have an overwhelming effect at first reading. Some readers, however, if honest, will admit with more or less diffidence that the play leaves them cold. That may be a final judgment, or it may yield to admiring enjoyment on repeated reading. Set off from average books, even from good books, classics require an attention all their own. Not all of them are necessarily well written in the literary sense of the word. The author's intent, his line of thought, his placing of values need careful scrutiny. A superficial perusal may be misleading and a bar to enjoyment. It may conceal our having brought the author down to our level of understanding instead of working ourselves up to his. The recognized classics have stimulated a huge amount of interpretative writing— none more than the Bible—and new points are constantly discovered. Interpretation (i.e., assimilation of a work by our minds) is an act that a few master intuitively, but most of us have to acquire. In bygone days, education included a painstaking, and often painful, study of languages, especially Latin, and taught the pupil to cope with what was difficult and strange. With the decline of the thorough study of languages, dead or alive, this educational help can no longer be counted on.

There is no contradiction in the contention that enjoyment of books is often coupled with work; in music and the visual arts, too, full enjoyment has to be earned. Better perhaps: the work done on a classic, and crowned by understanding it, can be enjoyable.

De gustibus non est disputandum said a scholastic adage—no arguing about tastes! Only a questionnaire could supply an answer to what readers ac-

tually do enjoy in medical classics; nevertheless, a few points suggest themselves. The enjoyment may act as a mental stimulus, refreshing the intellect, increasing the eagerness to go about one's own task and overcome the obstacles in its way. Depending on the content of the book in question, the gain can be more specific. Harvey's *De motu cordis* is a model for the development of a revolutionary scientific idea of great significance, incompatible with generally accepted doctrine. It can be read with a single-minded concentration on its thesis, that the blood circulates, and on Harvey's arguments, his methods, his diction only inasfar as they prove that thesis. But such an approach bereaves us of the pleasure of meeting the Harvey who was not just the discoverer of the circulation. Harvey had his predecessors; sooner or later somebody else would have given us this discovery. Sociologists have familiarized us with the idea that discoveries usually are made independently by several persons around the same time. But that is only true as far as "the discovery" is concerned. In reading *De motu cordis* we also learn the particulars of Harvey's approach to the discovery, what it meant to him, and the mental images he associated with it. We meet a particular mind, William Harvey's, trying to be as convincing as he knew how.

Claude Bernard was not the inventor of experimental medicine, yet a similar case can be made for his *Introduction to the Study of Experimental Medicine*, which gave it a philosophical exposition unmatched for breadth, lucidity of style, and persuasiveness stemming not only from logical coherence but also from the harmony between theory and the illuminating examples from his own research. There is much to criticize in Bernard's book, to mention only the lack of appreciation of the statistical method. But rarely, if ever, will medical classics represent the last word about a subject. "That living intellect that bred them," to use Milton's words, is the main thing. To meet that intellect is a privilege; it speaks to us and compels us to listen attentively to its every word.

As long as the reading of classics is felt to be a chore, it is a barren enterprise and had better be dropped, for it does neither the reader nor the books any good. It makes us rebellious and brings them into disrepute, heightening the barrier to a potential source of productive joy. The barrier need not be constitutional. Indifference or even hostility in early years may be dispelled by the discovery later in life of a book that strikes us as above the ordinary. We find sympathy where we thought to stand alone;

we find a keenness of observation that puts us to shame, a clear grasp of ideas behind primitive techniques, or an attitude toward medical duties that we cannot but admire. Instead of a chance discovery, a rational approach can also lead to classical works. At some time in our life, we realize that the basic concepts of medicine, which appeared so familiar, are more puzzling than complicated problems of molecular biology. Then we shall ask when and under what circumstances this concept was introduced and how it developed. That is the historical path. But even without tracing the steps through the past, we may interview the men who forged these concepts.

Our very familiarity with these concepts has prevented us from seeing the inherent assumptions. And so we rightly turn to authors who made the assumptions stand out because they were not yet conventionalized. We know more about "inborn errors of metabolism" and of "chemical individuality" than Archibald Garrod could possibly have known. But the ideas were his, and this accounts for the fact that interest in his book is very great today, much greater than it was during his lifetime.[5]

Whatever the occasion for a meeting with medical classics may be, the meeting becomes one between master and pupil. Not all of us have had the benefit of being taught by great medical teachers viva voce. The classics teach silently; they are teachers who will not answer every question, but who will help us to ask the right questions and to find the way by which we ourselves may work out the answers.

Yet reading such works is not without danger, for the light they shed may be dazzling. We may forget that in reality matters were not always as clear and unequivocal as the mind of the master makes them appear in his presentation. The logic of Henle's *On Miasms and Contagions* makes it hard to understand how his contemporaries could fail to accept the micro-organismic theory of infectious diseases. Only by immersing ourselves in the literature of the day do we realize how confusing the situation was.

Another, though minor, danger threatens even those who do not turn to the classics for historical understanding. Classics have to be read carefully because we expect their authors to have used their words carefully. Such reading, however, should not lead to contempt of what may be ephemeral yet important for doing "what lies clearly at hand." The words are Carlyle's, they are cited by Osler,[6] and I quote them to obviate the misunderstanding that there is bound to be antagonism between works of information and classics.

This is reminiscent of Osler's advice to a budding medical man: "Let the old men read new books; you read the journals and the old books."[7] Among the old books, Osler here specifically included the ancient Greeks, Sydenham, and Laennec—many, if not all, of whom would be considered classics. His subsequent remark "as a teacher you can never get *orientiert* without a knowledge of the Fathers, ancient and modern," indicates that the young man was a teacher, present or prospective. Altogether, it would seem that when speaking about the reading of medical classics, Osler exercised his role of medical educator. His activity in the United States belonged to the early period of medical reform; it ended about five years before the Flexner Report was issued and when the Johns Hopkins School of Medicine was still a new venture, barely twelve years old. The fight for better preparation of prospective medical students, for the academic character of medical schools, for a thorough scientific training of the physician and for his unremitting contact with scientific progress, formed the background of much of Osler's thought about "Books and Men," if we may use the title of this essay as denoting the theme dominant or touched upon in many others. Osler was pressing for good medical libraries and for good reading habits, in which he wisely included nonmedical works.

Many of Osler's desiderata have been fulfilled, though not always quite as he had envisaged them. The humanistic side of the education of physicians has not kept pace with their scientific training; on the other hand, medicine has moved closer to public health and to the social sciences. But the provision of good medical libraries is as little a matter of debate today as is the provision of good laboratories. If, in Osler's days, not enough important medical literature was housed, the question now is one of providing proper housing for the medical literature that is pouring out like a torrent.

With this question we are back where we started, only that our specific problem now is: Where do medical classics stand in this changed world? Do they belong with the active, informative books, or do they belong in the historical collection? Or should Osler's wish be heeded that there be "in each library a select company of the immortals set apart for special adoration"?[8] It would be simple to send all classics to the historical collection—if classics only had to be old and clearly separable from current life. And a chapel dedicated to the immortals could combine the

practical with the decorative—if only the selection of the immortals were a matter of easy agreement. As long as potential readers can find the books, their physical placement is a matter that concerns the librarian more than us, who ask whether classics have to be old and how a decision can be reached whether a book is immortal.

Books that have been read for many generations are said to have passed the test of time, a test that is unreliable for classics. One of the collections of Osler's essays from which I quoted carries the subtitle, "and Other Papers That Have Stood the Test of Time." Indeed, many of the essays are still read in Anglo-Saxon countries; I wonder how well they are known in other parts, even of our own Western civilization. Classics are bound to their respective civilizations. And there are essays by Osler that did not appeal to my students, required to read them. Indeed, I must admit that "A Way of Life," breathing, as it does, the optimistic spirit of 1913, does not easily fit into our benighted world. Characteristic as it may be for Osler's view of life, its prescriptions may be unconvincing and its prescriptive style may sound harsh to young ears today.[9]

Osler, too, was time-bound, and reading him and holding him up as an ideal should not become a substitute for emulating him. Where the humanities are concerned, it seems to me that what Galen said of Hippocrates could also be said of Osler: Most physicians admire him but do nothing to emulate him.[10] Emulating Osler means actually reading books (besides the informative ones) and forming one's own judgment about them, even at the peril of disagreement with the master.

My argument is not directed against the established classics as such, and the value of lists of "great books" for the schoolroom need not be denied, particularly if there is a competent teacher to select, to advise, and to comment. But there is a difference between books that should be read because our civilization has been formed by them and the advocacy of reading classics for their intrinsic value.

It is customary to refer to all ancient Greek and Roman authors as "classical," and in the historical sense this appellation is justifiable. But their intrinsic value may not warrant it. Galen's *Ars medica*, the *Microtegni* of the Middle Ages, is a great medical book that helped to form the outlook of medical men for hundreds of years; nevertheless, I would hesitate to offer it to a physician as a classic. One generation may well need a vaca-

tion from such classics as another may restore; "timelessness" and "immortality" should not be interpreted so rigidly as to deaden response and bar newcomers.

The opposite extreme to mere traditionalism is the devaluation of all classics by the cheapening of the very word. It is good that reprints, paperbacks, and anthologies put so many books within easy reach, thereby purging them of possible snob value. But the cheapening that derives from promiscuously naming all kinds of books "classics" is not good. It weakens our sensibility and creates disrespect for greatness. It is in line with the carelessness that calls any theoretical thought "philosophy" and misnames any knowledge "wisdom." As a result of such leveling we have no words left for real philosophy, real wisdom, and real classics—all of which are rare. Where everything is made to look like gold, real gold will not look different from anything else.

If the test of time cannot be relied on and everyday parlance is treacherous, on what shall we rely? On ourselves, I would suggest. If, upon reading a book (or an article) carefully and thoughtfully, having assured ourselves that enthusiasm has not made us read into it more than it demonstrably contains nor leave out parts because we cannot understand them, if then we judge the work to be exemplary of its kind, to set a standard, we shall have judged it a classic. *Exemplary* can refer to presentation, novelty of experience or interpretation, breadth of concept, coherence and consistency of thought, fullness and keenness of observation, ingenuity in experimentation, nobility of deed, or any other value that makes this work stand out even among good ones and tells us: This is the way to think, to write, or to act.

We shall have judged it a classic because, for us, individually, it has fulfilled the demands that can be made of a classic. If the book already stands in the alcove of the immortals, we shall have confirmed the judgment of those before us. In the case of elderly books or of active ones, others will or will not share our judgment, and the book will or will not gain the general approval we think it deserves. Official classification is the concern of educators, literary critics, bibliographers, and librarians. Classification becomes meaningless if classics are not read and either reconfirmed or suspended. And their being read depends on our capability, active or potential, to recognize their worth.

Two serious objections may be raised. To base the reading of classics

on subjective experience makes greatness relative, and only the absolute, it may be said, is binding. There is, perhaps, no better reply to this objection than the following words of Jacob Burckhardt, the author of *The Civilization of the Renaissance:*

> We take our departure from our dwarfishness, our confusion, and distraction. Greatness is what *we* are *not.* To the beetle in the grass, being nothing but a beetle, even a hazel-bush (provided it takes any notice of it) may already loom very large. And yet we feel that the concept [of greatness] is indispensable and that we must not allow it to be taken away from us. Only it will remain relative; we cannot hope to reach an absolute concept.[11]

Eschewing a systematic discussion and admitting all kinds of dangers, Burckhardt insisted that the feeling for greatness, though relative, is yet inescapable. Thereby he saved this concept from being lost in a discursive debate over its essence. This seems important in our time, when greatness is often attributed unreflectingly, sometimes even irrationally, or its existence denied altogether because it does not easily fit into a statistical world of quantities.

The second objection questions whether upholding the standard of the classics is not tantamount to perfectionism. Will it not induce us to hide our own light under a bushel? I agree that perfectionism is sterile when it becomes an excuse for never finishing a piece of work. But to quote a "classical" author, Propertius: "in magnis et voluisse sat est" [in mighty enterprises it is enough even to have willed success].[12] There is nothing wrong with trying to do one's best, and our libraries amply prove that our times do not suffer from a surplus of bushels.

Like all classics, medical classics, too, lead a pitiful existence if, like gods on Olympus, they are separated from us. In libraries they must not be shelved out of sight, even though their choice may demand a constant exchange of opinion within the medical community. For us as individuals they must be like the Roman *lares* and *penates,* gods of the household.

NOTES

This chapter is dedicated to my old friend Samuel X Radbill, M.D., a lover and reader of medical classics.

1. John Milton, *Complete Prose Works* (New Haven: Yale University Press, 1959), 492.

2. C. G. Benjamin, "Soaring prices and sinking sales of science monographs," *Science* 183 (1974): 282–84.

3. By definition every book has a literary aspect, however small a role that aspect may play.

4. Claude Bernard, *An Introduction to the Study of Experimental Medicine,* trans. Henry Copley Greene (reprint, Henry Schuman, 1949), 43.

5. Barton Childs, "Sir Archibald Garrod's conception of chemical individuality: a modern appreciation," *N Engl J Med* 282 (1970): 71–77.

6. "Our main business is not to see what lies dimly at a distance, but to do what lies clearly at hand." William Osler, "A way of life," in *A Way of Life and Other Selected Writings of Sir William Osler* (reprint New York: Dover Publications, 1958), 240.

7. William Osler, "Internal medicine as a vocation," in *Aequanimitas and Other Papers That Have Stood the Test of Time* (New York: W. W. Norton, 1963), 84.

8. Osler, "Books and men," in *A Way of Life,* 38.

9. My experience goes back many years, and generations of students are notoriously apt to change their opinions. Dr. James A. Knight's experience seems to be different from mine; see his essay on "The Relevance of Osler for Today's Humanity-Oriented Medical Student," in *Humanism in Medicine,* ed. John P. McGovern and Chester R. Burns (Springfield, Ill.: Charles C Thomas, 1973), 95.

10. Galen, *Quod optimus medicus sit quoque philosophus,* chap. 1.

11. Jacob Burckhardt, *Weltgeschichtliche Betrachtungen,* ed. Werner Kaegi (Bern: Hallwag, 1941), 314 (italics in the text). Burckhardt was not displaying false modesty: he referred to the encounter with what makes us feel small.

12. Propertius, *Elegies* 2.10.6, trans H. E. Butler (Loeb Classical Library), 91 (slightly modified).

CHAPTER 14

The Study of the History of Medicine

In recent years, much has been written and more has been said about the teaching of medical history. I do not wish to belabor that topic anew—especially since last year the history of medicine was made a required subject at our medical school. Instead I would like to discuss with you the study of medical history. I shall assume that one *wishes* to study the history of medicine—something a teacher cannot take for granted among his pupils. I shall merely raise the question: in what spirit and where should this be done?

Before this question can be approached, we have to clarify whom we mean by "one." There are many potential students, and the task is different for each of them. There are, for instance, the novelist, the playwright—and even the producer of historical films, whom I would ardently like to see a student of medical history. Then, perhaps, the day would come when the doctor of the eighteenth century would no longer pull a stethoscope out of his pocket and, having performed this anachronistic feat, proceed to apply the earpiece to the patient's chest. Such studies could be confined to external appearances and need be little concerned with a deeper understanding of medical conditions and science, which can be left to the historian interested in medicine and to the physician interested in history. It is to these two that I would like to devote the rest of our discussion.

Let us take the historian first. By definition a historian is a person who wants to know the past. He will ask what was done in medicine, by whom, for what purpose, and how? He will soon perceive that these questions

Originally published in *Bulletin of the Johns Hopkins Hospital* 104, no. 3 (March 1959). © The Johns Hopkins University Press.

cannot be answered without a much wider setting. For instance, the healing art in this country, some two hundred years ago, was not as sharply divided into the separate domains of physicians and surgeons as it was in Europe. Under the conditions of a thinly settled continent, this obviously could not be the case, and John Morgan failed when, in 1765, he wished to introduce the separation.

As this very simple example shows, the historian must know the relationship of medicine to other conditions of the time. This will make him inclined to break up medicine into the areas of different civilizations and periods. For the older civilizations a thorough study presupposes linguistic knowledge, which is a study in itself. Whoever works on Egyptian or Babylonian medicine without knowing hieroglyphic and cuneiform scripts does so at the peril of dilettantism. For many periods, a knowledge of the times is more important than a technical knowledge of medicine. Dr. Edelstein's work on Hippocratic medicine is an excellent example. While most physicians were fascinated by the medical meaning of prognosis, which predominates in the Hippocratic writings, and tried to read a deep scientific significance into it, Dr. Edelstein found a social motivation. The ancient doctor had to create his clientele and to impress it by announcing the past, present, and future conditions of the patient, and he had to exonerate himself in advance in hopeless cases. For more recent times, I might cite Professor Shryock's *Development of Modern Medicine* as an outstanding example of the successful approach by the general historian.[1]

I by no means believe that the historian should concern himself only with the background (social, economic, cultural, and intellectual) of medicine. In our present context I use the word *historian* to denote anybody who wishes to know how things in medicine were. Obviously, that relates to medical and scientific content, no less than to social environment. What Galen knew about anatomy, what discoveries William Beaumont made about gastric function, what the relative incidence of diseases was and how they were treated at the Johns Hopkins Hospital during the first years of its existence—all these questions belong to the realm of the historian. But whereas a study of the background necessitates, above all, a knowledge of existing general conditions, questions like the above depend on familiarity with medicine and its sciences. This has given rise to a serious problem. A historian who makes medicine the object of his study should know

general history and he should know medicine; how can these two be combined?

It is interesting to see the difference between the solutions that were offered formerly and those offered now. Medical history, until thirty years ago, was undisputedly in the hands of medically trained persons. To be sure, there were some philologists and a few historians who occupied themselves with editions of old medical texts and with ancient or medieval medicine. But they were not medical historians like Sudhoff, Diepgen, Sigerist, Neuburger, Garrison, Singer, Castiglioni, all of whom were doctors of medicine. In those days one became a historian of medicine as one becomes, for instance, a pathologist today. A knowledge of Latin and Greek and of some modern European languages was taken for granted. Either in connection with the thesis required in most European countries for the M.D. degree or in some other way later on, a connection was established with a professor of medical history. Paid positions being extremely few, much had to be done privately. But with or without a job, one tried to learn such historical methods as were needed for one's research. A humanistic education, a course during student days, supervised work on the thesis, and much private study constituted the main elements in the training of so-called historians of medicine. The result was not altogether satisfactory in every respect; many mistakes were made which a regular student of history is taught to avoid in his first year. But a tradition was established and maintained: A medical historian was able to understand and, if necessary, teach the history of medicine as a whole (i.e., of all times and in all aspects). He was supposed to be able to contribute to the understanding of the development that extends from Hippocrates to our own days, even if he considered some particular period the province of his more sustained research.

This relatively simple scheme has been upset by several factors: In this country the knowledge of foreign languages in general, and of the classical languages in particular, has declined to a degree that for the average M.D. all but excludes a serious occupation with medicine outside the Anglo-Saxon civilization. Together with this loss has come an increasing inability to enter into the spirit of a civilization fundamentally different from ours. The boy who labored over Homer and Cicero for years knew at least one thing: it had all been so very different! Our loss of feeling

for such differences has as a result that the philologist and general historian are now taking over what was once done by physicians. Moreover, historical studies themselves have advanced far beyond their erstwhile simplicity. Not only have they acquired a strong sociological element, but they have also extended outside their traditional fields. The social historian encounters medicine as a very important phenomenon: sickness, mortality, and the institutions for their prevention and cure are social factors, to be studied and understood. The medical student obviously cannot cope with the trained young historian—even if he wished to do so. Gradually, the serious study of medical history is being transferred to the general historian. But the general historian is "general" for his particular period only; he is not prepared to deal with medicine of all times. Moreover, he encounters a serious limitation in the subject matter. Thus, a division of labor has been suggested: Let the historian look after the social and cultural development of medicine, and let the medical man study the technical developments. I confess that this solution reminds me of Solomon's judgment: it will lead to the death of the child.

So far we have considered the historian who wants to know how things once were in medicine. We have used the term *historian* merely from the point of view of interest—not of schooling. Karl Sudhoff, for instance, was a trained medical man who for many years had engaged in the practice of medicine. But he investigated medicine of the past, particularly of the Middle Ages, as a medievalist would. He tried to find new texts; he deciphered, explained, and published them. For the rest, he was little concerned with their bearing on present or future medicine. In short, he was not a physician who turned to history because he expected from it elucidation of his practical and scientific work.

Lest we overlook the obvious, we must remember that at all times there has been the sheer enjoyment of good stories and anecdotes. This element must be rather strong, judging from the use made of it by advertising. Pamphlets, and even journals, brought out by drug houses are replete with articles and illustrations relating to medical history. This would not be so if history did not hold an attraction for laymen and physicians alike. Such articles vary from excellent historical accounts written by good historians to vulgarizations abounding in mistakes. To be sure, all history has a romantic aspect, which may be the element that makes it humanly attractive. But the publications I am now speaking of are not always history.

Sometimes they respond to the legitimate desire for entertainment or to the not so legitimate desire for introduction to the magic of miracle drugs and the working of the modern shaman called "the man in white."

No serious study is involved on the part of the person who wishes to be entertained. But how about the scientists and physicians who wish to consult the past for their immediate concerns? It can be said that the number of scientists who expect from history something other than mere knowledge of the past is growing. There exists, today, an increasing need for orientation in the meaning of fundamental concepts. Unless I am mistaken, physicists have taken the lead, but biologists and physicians are following. Before I came to the United States in 1932, I had done some work in the history of the concept of disease. But in 1932 I did not feel our climate here propitious for that kind of research. By now, the situation has changed. Concept of disease is one example among many where an elucidation of present thinking is expected from the tracing of the historical development. But—and here lies the difference from history as such—the past is studied only insofar as it promises to be helpful. The scientist does not want history; he wants something of history. As an example I may quote J. R. Baker, who accompanied an admirably painstaking reexamination of the history of the cell theory by the following remark: "Though I have great respect for the history of science, yet my main purpose in this series of papers has not been to write history, but to use a mainly historical method to establish what I believe to be an important truth about living organisms."[2] And when we turn to the doctor who asks of history to keep alive what is best in a tradition and to let the great masters of the art speak, we encounter a similar situation. Since I do not practice medicine, let alone surgery, I must be forgiven if I try to project myself into the mind of a surgeon. But conversations with surgeons and other practitioners have led me to believe that, to many of them, thoughts, manners, and even mannerisms of the great teachers have a deep meaning. Halsted is not just a name connected with certain operations. It is a way of approaching surgical problems and dealing with them. Other great surgeons had different ways. In spite of the fact that medicine has become "scientific" and much has become standardized, it involves a good deal of art. Art is learned by apprenticeship from example, and history is to keep alive the great examples.

I express these thoughts with some diffidence, but I have to enter upon them because they concern the study of the history of medicine. To put

it bluntly, they lead to the question of whether the study of the history of medicine had not better be left where medicine is taught and practiced.

This question is not an idle one, and there is good reason to bring it up at our Medical History Club. Until some twenty-five years ago, this club met here in the hospital where it had been founded in 1890 by Osler, Welch, and Kelly. And until about the same time, the *Bulletin of the Johns Hopkins Hospital* carried a considerable number of historical articles, particularly those emanating from the papers presented before this club by students, members of the faculty, or invited guests. Then the *Bulletin of the History of Medicine* was founded as an independent organ, and for many years our meetings were held at the Institute of the History of Medicine. Thus, a good deal of the study of medical history left the medical departments and the hospital and migrated to the professional medical historians. The opinion has been voiced that a good deal of interest in the study of medical history left the school and the hospital at the same time.

I wish I could simply deny this. But I have to admit that to some extent I agree with this opinion. Whenever I go through the old volumes of the bulletin of the hospital, I am impressed by the excellence of so many of the historical papers. I must even admit that our students have ceased to produce the kind of historical papers that their elders apparently were able to deliver. And the fact that we are meeting in the hospital tonight shows that the officers of this club, several years ago, recognized the necessity of dividing our meetings between the hospital and the institute.

We thus seem to be confronted by a dilemma. On the one side, we have medical history as studied by the historian, with its ever increasing demands upon historical knowledge, paralleled by an ever decreasing mastery of historical tools on the part of medical men. On the other, we have the vital interest of the physician and medical scientist, who ask of history what the historian is not necessarily willing or able to give: the link between the present and that which did not die with the past. Does this dilemma require a solution? and if so, where is the solution to be found?

I believe that the dilemma needs a solution and that this solution will not be found in either of the two extremes. If the history of medicine were to become a mere branch of history, it would not cease to be of interest to medicine. Let us not forget that medicine itself is changing. The time is not very far past when medicine was the exclusive domain of doc-

tors of medicine. But in the broad sense, this is no longer the case. The licensed practitioner of medicine is on the way to becoming a specialist within the wider social structure that we call medicine. By *specialist* I mean the man who is in direct contact with patients, while an army of other specialists, mainly doctors of philosophy or of science, works in laboratories, libraries, offices, and public health agencies. This specialist role is enhanced in this country by the limitation of the number of M.D.s turned out by our medical schools. Doctors of medicine are so badly needed in their practical functions that few can be spared for other work, be it in the basic sciences or elsewhere. At any rate, the changing character of medicine will fortify the demand for historical analysis. Of this I feel sure. However, I believe that the historian who is concerned only with a knowledge of the past will not be able to answer all questions put to him. It is hardly his business to evaluate what was done formerly in comparison to what is being done today. Especially when questions of medical science or practice are concerned, the historian tends to be overawed by the achievements of modern medicine and to use them as an absolute measuring stick or to be satisfied with bringing out the fundamental differences between then and now. He will not easily be inclined to use history critically, yet constructively, except in the region of social conditions.

On the other hand it appears equally doubtful that we can go back to the time of Osler and keep historical studies a mere adjunct of the physiologist, internist, or surgeon. In making this statement I would like to head off a misunderstanding. I do not wish to see such studies weakened, let alone taken away, from the departments. On the contrary, I would do everything I could to encourage them and cooperate with them. The active medical scientist and practitioner has a direct, vital approach to his subject and a grasp on the more recent developments in his field that a so-called medical historian will not be able to emulate. It is here in the departments that our students must gain the feeling that the historical approach is the concern of all of us. Here are the grass roots of the history of medicine as a part of medicine. If these roots were to die, then as I said before, the history of medicine would not disappear, but it would become a subdivision of general history, looking upon medicine as an interesting object, without having a share in it. In saying that we cannot go back to the days of Osler, I meant that these studies by the medical man can no longer be carried on in isolation from the general study of history.

I do not know how far ideas like these were in the mind of Dr. Welch when he planned the Institute of the History of Medicine some thirty years ago. But it may not have been chance that led the first head of our pathology department to become the first director of the Institute of the History of Medicine. Within the present context, pathology and medical history seem to have much in common.

Before the days of Rokitansky and Virchow, pathology hardly existed as a specialty.[3] The clinician performed the autopsies on his cases, possibly helped by a prosector. It was, and still is, said that something was lost when the professor of pathology took over and that it would be a good thing to compel the clinician to go back to the autopsy table. Even if that were true, pathology was bound to become independent. It had to find its connection with general biology as a natural science. This led to Virchow's *Cellular Pathology*, published just a hundred years ago, and to pathology as an experimental science. It did not mean that pathology lost its autonomy. Today, every medical man studies pathology with a specialist. But the medical man also remains somewhat of a pathologist, regardless of where his particular field of interest lies. Indeed, he may know more of the pathology of a particular organ or disease with which he is concerned than does the professional pathologist. It would be just as impossible to take pathology out of the departments as it would be foolish to drive it back into the departments. Somewhere, pathology must be studied as a science, even though the pathologist may not be able to compete with all specialists in their particular fields.

Replace biology by history, and the parallel with medical history will be clear. I do not wish to force the analogy; it has many weaknesses and was intended only to indicate the place of the study of medical history as a discipline. This study will have to be pursued in close contact with the general historian as well as the medical scientist and practitioner. This means that a department of the history of medicine should have room for those who wish to know how things once were and for those who are pursuing historically an actual medical problem. The study of the history of medicine, even toward professional status, should not be made dependent on the degree the student holds. The general historian with his Ph.D. degree or candidacy for such a degree and the doctor of medicine or medical student, each has his advantages and handicaps. A department of the history of medicine should be prepared to help either of the two

in overcoming obstacles. As far as the medical man is concerned, this is best done by apprenticeship. Those of our medical students who have serious historical interests should be encouraged to try doing historical research work. Medical graduates may even study toward a Ph.D. degree, though I would recommend such a formal path to professional aspirants only. If a department has medical men working in it, their very presence will facilitate the path for the general historian or sociologist. In contact with them, he may learn what books alone cannot supply, an impression of the spirit of medicine. The position of the department within the medical school may then also enable him to establish contact with other medical departments. It goes without saying that his contribution in showing the medical student the methods of historical research can be very great.

All such arrangements are viewed from one central point: the pursuit of medical history as a subject combining in itself historical and medical interests. It seems to me that this combination of interests should characterize the medical historian, in distinction from the pure historian dealing with medicine and from the active medical scientist using history. That the borderlines on both sides will not be sharply drawn goes without saying. If I speak of combination of interests, I do not have in mind their mere addition. Elsewhere, I have tried to show how an organic union between historical and scientific research might be achieved.[4] Tonight I am dealing with *study*, which is more comprehensive than *research* and concerns a wider audience.

To sum up: the history of medicine can be studied as mere history; it can also be approached with a medical and scientific interest. Both approaches have their right to existence, both can lead to valuable results, and both should be cultivated. But beyond either of them there should be an organic synthesis, which we may call the history of medicine per se. The existence of such a discipline justifies the existence of independent departments of the history of medicine. Here at Johns Hopkins we are privileged in having an institute with excellent physical facilities, situated in a great medical library; we are linked with a philosophical faculty with a strong humanistic orientation, and we are part of a medical school famous throughout the world. Such a privilege imposes a duty: the duty to see to it that the spirit of the place and the work done in it be equal to the task and worthy of the place and its great tradition.

NOTES

This chapter began as a paper read before the Johns Hopkins Medical History Club, November 24, 1958.

1. L. Edelstein, *Peri aerōn und die Sammlung der hippokratischen Schriften* (Berlin: Weidmann, 1931); R. H. Shryock, *The Development of Modern Medicine* (New York: Alfred A. Knopf, 1947).

2. J. R. Baker. "The cell theory: part V," *Q J Microsc Sci* 96, 3d series (1955): 450.

3. In a conversation with Professor Shryock, I was reminded of the significance of pathology's early struggle for independence.

4. O. Temkin, "Scientific medicine and historical research" (in press).

CHAPTER 15

Wunderlich versus Haeser
A Controversy over Medical History

OWSEI TEMKIN AND C. LILIAN TEMKIN

In 1840 Heinrich Haeser (1811–85)[1] began editing a new medical journal, the *Archiv für die gesammte Medicin*. A prospectus had been circulated,[2] but the first issue did not carry any programmatic article. Instead, a preface, dated March 1841, appeared with the fourth issue, with which the first volume was completed. This preface set the journal the task of uniting men of science in a common activity: "Far removed from the indifference, devoid of strength or savour, of so many of its sister journals, the young periodical should take pride in the decisiveness with which it took sides."[3] It was to be on the side of progress, and the columns of the journal were to be open to anything offered in this spirit. The editor was particularly grateful to the contributors of historical articles, which had had no journalistic outlet since the demise of Hecker's *Annalen* but which would now give lasting value to his journal.[4]

In looking at the historical articles of the volume, we find the following situation. The leading article was by A. W. Henschel on Hippocrates.[5] He gives a very idealistic view: Hippocrates becomes comparable with Socrates only. Hippocratic medicine must not be measured by reference to science. It represents the artistic spirit that, according to Henschel, medicine badly needs. Immediately following is an article by Haeser himself on the question: "With what right is Paracelsus called the reformer of Medicine?" The answer is negative: Paracelsus was not the Reformer of medicine. "To be conscious of his proper task, and yet to miss its solution entirely by the most daring transgression of his own principles—just

Originally published in *Bulletin of the History of Medicine* 32, no. 4 (March–April 1958). © The Johns Hopkins University Press.

that is the peculiarity of this remarkable mind."[6] Haeser recognizes the excellence of some of Paracelsus's remarks but warns against taking them as proof of a reformation of medicine. The last major historical article of the volume is by Choulant: "Surgery in the earliest Middle Ages."[7] Choulant is still greatly valued today for his bibliographical work, and the article consists mainly of quotations from the Latin works of surgical authors of the thirteenth and fourteenth centuries. The historical articles are followed by a number of reviews of historical works: one by Rosenbaum (pp. 95–122) of the first volume of Littré's edition of Hippocrates, another by Rosenbaum (540–49) of Ritter's German translation of Celsus, one by Choulant (550–53) of Ermerin's *Anecdota medica graeca*, another by Choulant (553–56) of Sprenger's dissertation *De originibus medicinae arabicae sub khalifatu*, and two reviews by Haeser: on a book by Frari on the plague (560–64) and on a paper by Troxler, in which the latter stressed the necessity of studying the history of medicine and science in their national framework. This speech, intended for Troxler's countrymen, the Swiss, gave Haeser an opportunity to express himself on the necessity of having required courses on the history of medicine at the universities:

> Thus these considerations lead the author to a matter which is highly important not only for Switzerland, but for science in general: to the necessity for special professorial chairs for the history of medicine and the theory of national diseases. We should have plenty of reason to complain sadly if this matter needed any further demonstration after what has been cited by Troxler and still earlier by Hecker (namely in the Heidelberg Yearbook à propos the *Lettres médicales* of Dezeimeris). It is obvious that the reform in question, at present urgently demanded by the spirit of science, must come from the authorities. Not even the most distinguished teacher of history—and here it is just as much a question of external superiority of presentation as of thorough knowledge—will succeed in binding the students to his lecture-room as long as the examining body (for the most part for very valid internal reasons) does not concern itself with investigating the historical knowledge of the candidates, as long as the State, clinging to obsolete statutes, does not express in law its conviction of the necessity for the historical education of its physicians, and persists in its failure to recognize the infinite gain which it thereby forfeits.—How are we to define a situation in which at some universities medical students, like the others, are rightly obliged by law to take a complete course in philosophy and, in particular, to audit history of philosophy, while it is left to

their discretion to care, or not, to go to lectures on the history of medicine with the announcement of which some junior or senior lecturer is kind enough to complement the catalogue? Or is the majority of our young doctors perhaps capable of supplying the deficiency through their own studies? One needs to know our practising physicians in order to realize their candid and incredible ignorance of the history of their science.—May Germany be spared the shame of seeing this deficiency disappear only after Paris has again filled its momentarily orphaned chair of history; may it not experience the humiliation of receiving instruction in this matter from "barbarians." For, as far as we know, there are now in Europe, apart from Italy, only 2 chairs in the history of medicine: Berlin and—Petersburg![8]

While the second volume of Haeser's *Archiv* was in the process of publication, a new journal made its appearance: the *Archiv für physiologische Heilkunde,* edited by W. Roser and C. A. Wunderlich. The "Introduction" by the editors, "On the Defects of Present German Medicine and on the Necessity for a Decisive Scientific Direction in It," has become famous.[9] It was the battle cry, so to speak, of the new generation for a radical reform of medicine. The editors were very young: Wunderlich was born in 1815 and Roser in 1817. The tone was very aggressive, as was customary among the young leaders of the era of "medical reform."[10] The first article proper was written by Wunderlich. It dealt with "medical journalism,"[11] and here the criticism found its main target in Haeser's *Archiv für die gesammte Medicin.* Haeser's journal was accused of lacking "policy (*Tendenz*) and character."[12] Special attention was paid to the historical articles of the first volume, and in the following we give a translation of Wunderlich's interesting criticism:

The first number of the *Archiv* begins with two historical discussions; besides these, there follow several other essays of historical content, and these alone take up one-fifth of the first yearly volume. Convinced though we are of the importance of historical research properly undertaken, to us such a heaping of historical material in a journal addressed to the present seems at all events to be a blunder.

What we are, however, much more eager to express at the moment is the conviction that historical investigation of medicine has, to date, in no wise accomplished what it could accomplish, and that in it there is a predominant tendency which undoubtedly bears handsome witness to the re-

spectable studiousness of our countrymen, but which also betrays the lack of practical sense of which we have so often been accused. Häser's *Archiv* is a striking example of what we mean. Of the nine longer and shorter historical essays, two are concerned with Hippocrates; further, one with late Greek medicine, one with Celsus, one with Arabic medicine, one with surgery in the early Middle Ages, one with the Black Death in the fourteenth century, one with Paracelsus! One should not imagine that these essays are integrated and thereby embrace in monographic form the most important things in older medicine. For that they are much too concerned with often unimportant particulars and are not sufficiently instructive. No, chance brought them together, or rather, they were instigated by the widespread and inherited preference of our people to devote themselves to things which are as remote as possible. They have sprung from that honorable but unhappy antiquarianism (*Antiquitätengeist*) which, with the sincerest and most thorough earnestness chooses trifles as the object of its investigation, if only they are ancient, unfortunately overlooking in the process the great and the obvious which would make us feel at home in ancient Hellas and Rome, and would thereby suffer us to forget that we are Germans and that we live in the nineteenth century. We respect every kind of endeavor, but it is prejudicial to a practical science like medicine, if its historical scholars pay homage to a one-sided antiquarianism (*Alterthümlerei*). And one-sided it is, for in cultivating the old, what is nearest to us is neglected. We have monographs about the Greeks and the Scholastics, about Paracelsus and Van Helmont; the immediate past remains almost unnoticed. Häser's *Archiv*, which gives us eight articles on ancient and medieval medicine, gave only 1½ pages to more recent historical events—and in these it discusses Swiss national medicine! By this passion for the foreign, for the antiquated, and for unedifying specialties, the practicing physician is methodically estranged from historical study. What does he care about hearing of all the fantastic ideas with which one could impress the public in Paracelsus's time? What does he care about the Black Death or St. Anthony's fire? Of what concern is it to him whether people were syphilitic at the time of Celsus, or not? It is true, these are all very interesting questions, but interest in them is limited to an extremely small circle. Considering the indifference of the large medical public to historical investigations, history should not seek to impose itself upon the quite otherwise occupied physician with the impressiveness of all its unapproachable greatness, in all its learned majesty. It must first show him that it could be useful to him; it must first recommend itself to his attention by helping him to understand the time

in which he lives and to solve the problems that plague him in the present. Some will then not fail to penetrate more deeply into wider and more exact research.—You complain of the unscholarly temper of your colleagues who only yawn over the learned results of your night-watches. Blame first yourselves that you have not understood how to tickle the palate on which you urge an unaccustomed dish!—One should not imagine that a knowledge of the spirit of Hippocrates or Paracelsus is transmitted by way of a short journal article. Such an article is only understandable to one who is already at home in those times and circumstances and already familiar with the writings of those ancient physicians. In spite of all efforts to be objective, to the uninitiated such an abstract discussion is bound to look like the subjective opinion of the historian. And if he reads ten similar discussions of the same thing, he will still be unable to use them as a basis for forming an independent judgment.

Our purpose would be greatly misunderstood, if one were to believe that herewith we wish the study of old and ancient writers limited. No! We simply demand that more modern times be considered equally or rather more extensively. Haller, Zimmermann, Bordeu, the School of Montpellier, French surgery in the eighteenth century, Stoll, Cullen, Hunter, Bichat, Dupuytren, Rasori, Broussais, Laennec, Reil, Autenrieth, the German school of natural philosophy, etc., these are historical phenomena of which the ordinary doctor, and even many an author, has a strange and very incomplete picture. And yet the good and the bad sides of our present eclecticism are largely composed of these elements. Historical investigations of this more recent past, if undertaken in the right spirit, will never fail to be a beacon to the present and will persuade even the most practical practitioner that his knowledge is mere piecework, if he does not know whence it comes and the manner of thought of his predecessors.

Still more useful than the monographic treatment of single periods and important people would be yet another kind of historical study, which is also rather neglected nowadays. Individual doctrines, specialties, therapeutic methods, should be subjected to historical investigation. In this way the genesis of many theories and medical maxims can be shown, and we can demonstrate how they have been allowed to attain the status or incontestable facts and incontrovertible rules. One will discover the weak roots of many a traditional assumption whose importance one today scarcely thinks it necessary to prove. One will realize from what errors they sprang, how they, nevertheless, were able, even after the destruction of their

foundations, to propagate themselves from generation to generation without acquiring solidity through gaining factual support. One will realize how powerful a motive in orthodox medicine is habituation to a certain point of view and manner of action and how very often the consolation that the whole world believed in something silenced the question as to the real proof. History of this persuasion will be little concerned with ferreting out the peculiarities of famous men and strange times; rather it will trace the origin of the ideas (*Ursprung der Ideen*) that govern the thought of today. It will reveal how science was formed. The illusory in theory and practice is bound to melt before it, and room can be made for a more thoughtful view of things.[13]

This attack appeared in time to allow Haeser to address an open letter to Wunderlich, dated January 12, 1842, the year of the second volume of his *Archiv*, to which we find it appended.[14] Though Haeser is on the defensive, his reply lacks neither skill nor a certain justification. He reminded Wunderlich that they were all still very young and that it might be advisable to have some positive achievements to show before trying a general reform. Here Haeser had the advantage of having the *Historisch-pathologische Untersuchungen* to his credit, a work that was eventually to merge into the third volume of his great *History of Medicine*. Yet, just in his own field, that of medical history, Haeser's defense is relatively brief:

You base further objection to the *Archiv* on the historical essays, examples of which by Henschel, Choulant, Rosenbaum, recently also by Hecker, and myself have appeared there. The accumulation of historical material, in which you include not only the actual essays but also several critiques of historical medicine, literature, etc., you call a blunder, since a journal should be for the present. I must confess that I have attempted to the extent of my powers, and shall in the future so attempt, to give my journal more than ephemeral significance, and that it was my belief that this end could be attained with the aid, among other things, of studies from the pens of our most renowned historical investigators. I am namely of the opinion that just by being so fortunate as to open the *Archiv* with Henschel's study of Hippocrates I have made its first pages much more valuable than if I had planned some ephemeral journal and had shown that it, like all the others, is of no consequence, and that I personally am the messiah who will deliver Israel from her ancient misery. As far as my own historical essays are concerned, I surrender them to you completely; or

rather, your judgment of them is a matter of complete indifference to me, since I am bound to assume your complete ignorance of the history of your science, if only because of the tone in which you, respected colleague, without having as yet accomplished anything of importance in your field, judge the accomplishments of men who have, for a longer or shorter period of time, been recognized as tried physicians and scholars. This is not the preferred tone usually attained by thorough historical studies, in the pursuit of which the small, often pitiful, thing called the *ego* tends to become smaller and smaller. If only to gain this modesty, respected colleague, I should like to recommend the historical studies to you. You would then be less outspoken in your judgment of work which you reject with a tone of superiority, without having even the slightest idea of its aims and attainments. I must break off here, for I feel a slight wave of anger mounting in me, not on account of your censure, but because of your wrongheadedness and your lack of respect for a thing that would be so healthy just for you![15]

We have not been able to find a reply from Wunderlich, and this particular controversy over medical history seems to have come to an end here. It must, of course, be seen within the framework of the medicohistorical situation of the time.[16] However, if we confine ourselves to the immediate issue, we can characterize the opposing positions as follows. Wunderlich insists on the need for medical history of direct utility: concentration upon the immediate past with which all medical men have to come to terms and tracing of the ideas that concern the current theory and practice of medicine. Wunderlich is right in criticizing the amount of space that Haeser assigned to medical history in a general medical journal. On this point, Haeser's reply is weak. On the other hand, the historical articles and reviews would have done credit to a journal of medical history. Haeser's misplaced resentment may well have been motivated by the conviction that the value of medical history is in its long-range perspective and not in a one-sided emphasis on the "ephemeral."

The controversy represents two points of view, expressed by two different personalities.[17] Haeser, though never entirely severed from practical medicine, is remembered as one of the great historians of medicine whose classic textbook is still used today.[18] Wunderlich, though he, too, wrote a general history of medicine,[19] is remembered as a great clinician. Although their controversy seems bound to a particular situation in the development

of German medicine, we thought it remarkable that so many of the slogans of today can be found in the excerpts we translated: history of "ideas," the need for actuality as against "antiquarianism," the need for required teaching of medical history. The debate was not for or against the study and teaching of medical history. Both protagonists expressed their high respect for it, and both taught it to their students.[20] Rather, the discussion related to what kind of medical history should be cultivated by and for medical men. The professional historian of medicine wanted to spread his discipline, to cultivate the whole field, and to find in scholarship a counterbalance to ephemeral views, even if his investigations led him far away from the problems of the day. The clinician demanded emphasis on just those historical issues that contributed directly to the orientation and understanding of the practicing physician.

More than a hundred years have passed, yet not only the slogans but the very problems are still ours. We may take one or the other side, or we may be inclined to think that there is room for both. Medicine is not a purely contemplative science; whatever is to be a part of medicine must, qua such, be able to contribute something useful. Yet it is equally true that usefulness is a weak ground for any discipline. No discipline can flourish without being rooted in the search for knowledge unhampered by pragmatic considerations. At any rate, the realization that, in a stormy period of German medicine, ideas were voiced which we are inclined to think the product of our own stormy time, may be of help to all of us. It shows at least that there are few situations that do not have a precedent in the past. And the excited tone of the old feud may cause us to go about our work more calmly, conscious that if the argument is old, only the quality of our work, whatever its direction, can have something valuable to add.

NOTES

1. Julius Pagel, "Haeser," *Allgemeine Deutsche Biographie* 50 (suppl.): 53–54.
2. Heinrich Haeser, "Vorwort," *Arch Gesammte Med* 1 (1841): v: "Was die Redaction bezweckte, als sie die grosse Zahl der medicinischen Journale um ein neues vermehrte, darüber hat sie sich in dem das Unternehmen einleitenden Prospekte ausgesprochen." We have not seen the prospectus.
3. Ibid.
4. Ibid., vii.
5. A. W. Henschel, "Hippokrates," *Arch Gesammte Med* 1 (1841): 1–25.

6. H. Haeser, "Mit welchem Rechte wird Paracelsus der Reformator der Medicin genannt?" *Arch Gesammte Med* 1 (1841): see 35.

7. Ludwig Choulant, "Die Chirurgie im frühesten Mittelalter," *Arch Gesammte Med* 1 (1841): 417–35.

8. H. Haeser, "Review of 'Umrisse zur Entwickelungsgeschichte der vater-ländischen Natur- und Lebenskunde, der besten Quelle für das Studium und die Praxis der Medicin' by Dr. J. P. V. Troxler, 1839," *Arch Gesammte Med* 1 (1841): 558 f.

9. W. Roser and C. A. Wunderlich, "Ueber die Mängel der heutigen deutschen Medicin und über die Nothwendigkeit einer entschieden wissenschaftlichen Richtung in derselben," *Arch Physiol Heilkunde* 1 (1842): i–xxx.

10. On this period see Erwin H. Ackerknecht, "Beiträge zur Geschichte der Medizinal-reform von 1848," *Sudhoffs Arch Geschichte Med* 25 (1932): 61–109, 113–83.

11. C. A. Wunderlich, "Die medicinische Journalistik," *Arch Physiol Heilkunde* 1 (1842): 1–42.

12. Ibid., 9.

13. Ibid., 10–13.

14. H. Haeser, "Sendschreiben des Redacteurs des Archivs für die gesammte Medicin an den Mitredacteur des Archivs für physiologische Heilkunde, Herrn Dr. Wunderlich in Tübingen," 8 pp. In the copy of volume 2 of Haeser's *Archiv* (Jena, 1842) belonging to the College of Physicians of Philadelphia, the above *Sendschreiben* is bound in at the end of the volume. We take this opportunity to thank the college and Dr. W. B. McDaniel, 2d, for having made the first two volumes of Haeser's *Archiv* available to us.

15. Ibid., 5–6.

16. Ludwig Edelstein, "Medical historiography in 1847," *Bull Hist Med* 21 (1947): 495–511; Edith Heischkel, "Die deutsche Medizingeschichtschreibung in der ersten Hälfte des 19. Jahrhunderts," *Klin Wochnschr* 12 (1933): 714–17.

17. Paul Diepgen, "Das Schicksal der deutschen Medizingeschichte im Zeitalter der Naturwissenschaften und ihre Aufgaben in der Gegenwart," *Deutsche Med Wochnschr* 2 (1934): 66–70; see 66 f., where Diepgen contrasts Haeser's and Wunderlich's attitudes. See also by the same author: *Geschichte der Medizin* (Berlin: De Gruyter, 1951) 2:217.

18. Heinrich Haeser, *Lehrbuch der Geschichte der Medicin und der epidemischen Krankheiten*, 3 vols. (Jena, 1882), Dritte Bearbeitung.

19. C. A. Wunderlich, *Geschichte der Medicin* (Stuttgart: Ebner and Seubert, 1859).

20. The subtitle of Wunderlich's *Geschichte der Medicin* reads "Vorlesungen gehalten zu Leipzig im Sommersemester, 1858."

CHAPTER 16

In Memory of Ludwig Edelstein

We are here to express our feeling of loss over the death of Ludwig Edelstein and to pay an early tribute to his memory. It is too soon yet to evaluate his work; we hardly know what he has left us for publication. While he was with us, he was a friend, a colleague, a teacher; we discussed with him weighty things and talked about ephemeral ones. He meant much to all of us, but he also meant something different to everyone. Looking backward, we want to collect our memories and try to see him once more before us.

Ludwig Edelstein was born in Berlin, Germany, on April 23, 1902, as the son of a businessman. Having received a classical education, he turned to the study of philosophy. As he once told me, he had tried science as an entry to philosophy, but this had been a short episode. Always "charmed" by ancient Greece, he found the right approach in classical philology, which he studied in Berlin and in Heidelberg. It was in Heidelberg that Otto Regenbogen was instrumental in introducing him to the Hippocratic works. In 1928 he married Emma J. Levy, and in 1929 he received the Ph.D. degree with his dissertation, *Peri Aerōn und die Sammlung der Hippokratischen Schriften*. The book meant a revolution in our concepts of ancient medicine; it also characterized the nature of Edelstein's research, his uncanny ability to approach an ancient text without prejudice, to let the text speak fully for itself, and to follow the consequences of his interpretation. Reinterpreting two words, namely, *to holon*, in Plato's *Phaedrus*, he transformed the alleged natural philosopher Hippocrates into a physician who insisted

Originally published in *Bulletin of the History of Medicine* 40, no. 1 (January–February 1966). © The Johns Hopkins University Press.

on knowledge of the body as a whole. Instead of trying to identify the genuine writings of Hippocrates in the collection that bears his name, Edelstein declared our inability to make such an identification. He even suggested that none of the so-called Hippocratic books corresponded to the teachings of Hippocrates as reported by Aristotle's pupil Menon.

Far from bringing him recognition, the book roused the antagonism of many classical scholars and historians of medicine. For many years this fate accompanied Edelstein's medicohistorical publications: on the history of dissection in antiquity, on the ancient medical sects, even on Asclepius. Edelstein was a man of great intellectual courage who followed the trend of his thought regardless of the consequences involved. He would take endless time and pains over his work. When he praised Arthur Lovejoy's dedication to details, he proclaimed his own ideal. This ideal was not merely *akribeia*, the exactness demanded of the philologist. Research had to lead to truth; if it was satisfied with less, "it made no sense." This expression he often used when confronted with work that did not keep its promise, be it that the author was incompetent, be it that a competent author had yielded to the temptations of intellectual laziness or opportunism. Though in his scientific writings Edelstein would contradict any person whom he thought to be wrong, he would do so by objective argument only, free from personal polemic. Thus he encountered personal enmities and the resistance of those who were not pleased to see their own opinions controverted in a spirit of objectivity that seemed to slight their authority.

This was the state of affairs from the time of Edelstein's graduation until about the end of World War II, the period during which he had to struggle for recognition. Personal and political events combined to make the struggle a difficult one. His father, to whom he was deeply attached, was very ill, and during this illness and after his father's death, Edelstein tried to keep the paternal business going, at a time of increasing economic depression. His plan to combine business activity and scholarship was bound to fail, though loyalty to his father's memory and to the old employees of the firm made him resist necessary dissociation as long as he could. Professor Diepgen offered him a position in the Berlin Institute of the History of Medicine and Science, and the University of Berlin put him in charge of a course. But in early 1933, when Hitler came to power, even this arrangement broke down. He and his wife (who had just received

her Ph.D. degree in classics) thought of working in Italian libraries. The generosity of a private individual made this plan possible, and in September 1933, the Edelsteins found themselves in Rome, where they stayed for approximately a year. Then they came to Baltimore, where Dr. Sigerist had obtained the funds from philanthropic societies to secure Edelstein's salary as an associate in the history of medicine until he could be absorbed into the regular budget of the department. Edelstein remained with the institute for thirteen years, obtaining the rank of associate professor in 1939. During this period there appeared his book on the Hippocratic Oath and the monumental two volumes on Asclepius on which he and his wife had worked together.

He also ventured into more recent periods when he doubted the legend of the book-shy Sydenham or brought William Osler in contact with William James and pragmatism. Edelstein had ample time for his own research. The teaching load and other duties were light and allowed him the necessary leisure to read widely in classical literature. He engaged upon the collection of the fragments of Posidonius, and it was largely upon his initiative that Carnegie Fellowships in the History of Graeco-Roman Science were established, bringing Dr. I. E. Drabkin and Dr. G. R. Thompson to the institute. He formed lasting friendships in Baltimore, and he found devoted pupils among the medical students, to whom he even offered a course in philosophy that was received enthusiastically. The History of Ideas Club made him its president in 1944–45, after he had served as its secretary for the preceding four years. He was devoted to the club's main founder, Arthur O. Lovejoy, and with D. C. Allen and George Boas he helped to bring out a collection of Lovejoy's *Essays in the History of Ideas.* And yet, he had not found the right place, in spite of his connections with the Philosophical Faculty. A department in a medical school could not offer him an adequate basis for his activity and the prospects which he, as a philologist and philosopher, needed and claimed. Thus, in 1947, when Dr. Sigerist left the United States and Edelstein's own future in Baltimore appeared bleak, he accepted a position in classics at the University of Washington in Seattle. A year later he found himself in Berkeley at the University of California as professor of Greek.

With his departure from Baltimore, a new phase began in Edelstein's life. Contributions to ancient philosophy might now be expected to flow more freely. Yet his productive work during the next few years, naturally

limited by the necessity of adapting himself to new surroundings and new duties, soon was overshadowed by the controversy over the oath, which began to stir the University of California in 1950. Edelstein belonged to the small group of nonsigners who finally won their case against the university before the Supreme Court of California. He was not willing to bow to a political demand which, in his opinion, the authorities of the university had no right to make a condition of employment and which was prejudicial to the integrity of the academic teacher. A convinced democrat, he was as outspoken in his rejection of extremism from the left as he was firm in his resistance to illiberal moves in the name of fighting such extremism; and, indeed, he could be outspoken enough whenever he believed that fairness or principles demanded it. Those were difficult years for him and his wife, but their steadfastness not only helped his cause, it also deeply impressed people far beyond the circle of his personal friends. In 1951 the Johns Hopkins University, through President Bronk, invited him as visiting professor of philosophy and in 1952 appointed him professor of humanistic studies. In the same year he reestablished his formal ties with the Institute of the History of Medicine, with which he remained affiliated until the time of his death.

For the Edelsteins, the return to Johns Hopkins was "a homecoming," as they repeatedly called it. The resistance with which he had so frequently met was giving way to an ever increasing respect for his opinions and for the man. He served as a Fulbright Lecturer at Oxford University in 1953; he was elected a member of the American Philosophical Society in 1954 and to honorary membership in Phi Beta Kappa in 1955. In the same year he gave the first William Osler Memorial Lecture at McGill. He was a member of the Institute of Advanced Study in 1959–60 and Garrison Lecturer of the American Association for the History of Medicine in 1960.

Though external honors were never sought, they yet gave him the satisfaction of seeing his work recognized. In those years he was able to derive pleasure from them, and he had preserved a truly childlike happiness in celebrating his birthday. However, all this changed in 1958 when his wife died. He took the immediate event with almost frightening self-control, reciting a passage from Plato at the funeral with unfaltering voice. Afterward he retired into a semiseclusion that lasted for years and caused great anxiety to his friends. In 1960 he left Baltimore to go to the Rockefeller Institute, though he preserved a part-time affiliation with Johns Hopkins

as professor of classical philosophy and as lecturer in the history of medicine. His physical health was far from satisfactory, but he had regained openness to the world around him and even some degree of cheerfulness when he died on August 16, 1965, in his home in New York.

→→ • ←←

Edelstein's was too deep and too complex a nature to permit any facile characterization. All I can do is to gather some external manifestations that to me seem to have a common source. The effect of his wife's death, which came close to a catastrophe, was due to the love and reverence he bore her and the intimacy of their life and work. It was also anchored in the seriousness with which Edelstein considered all life and death. Years later, in a critical mood, he would write: "Even at the death of those whom we love, we shrink from dying a little ourselves, from withdrawing from the world of the living."[1] The death of any friend, even of a mere acquaintance, shocked him deeply. Edelstein did not subscribe to Seneca's *humanius est deridere vitam quam deplorare.* Never given to levity, he always looked for significance in the events of his own life and the events and lives around him. The interpreter of classical texts was also an interpreter of men and their actions as he saw them. Again and again he impressed me with the importance that he attributed to the happenings in his proximity, to the people he knew and to the institutions he knew. There seemed to exist a parallel between his interpretation of texts, often mere fragments, which led him to far-reaching conclusions, and his interpretation of life, often mere incidents, which led him to an insight into affairs at large.

Edelstein's strong attachment to those he knew showed itself in the obligation he felt to take a dead friend's legacy upon himself. He made possible the posthumous publication of the Noguchi Lectures on Hindu Medicine of Heinrich Zimmer, where he edited what existed in manuscript form and added a long preface that carried him into a field not primarily his own. He brought out a collection of essays by his friend, the philosopher Erich Frank, together with an appreciation of Frank's work. After the death of Sigerist, he consented to edit the second volume of Sigerist's *History of Medicine,* a work in which he was helped by Miriam Drabkin. Finally, after his wife had died, he spent years on making possible the publication of a book on which she had been engaged.

Edelstein's piety toward his departed friends was in harmony with his

conservative turn of mind. In speaking of his conservatism, I do not, of course, take the term in any narrow political connotation, nor do I imply that Edelstein favored a hanging on to vested interests. Nevertheless, much in modern literature and art was decidedly not to his taste. All the greater was his dedication to what had once found his approval. In his younger years, he had been an ardent pianist, and he loved classical music, above all Beethoven. He was fond of classical German literature, which he knew well, as his essay on Wieland evinces. For Jakob Burckhardt he held a great and abiding admiration, an admiration that, I feel, included not only the man but also the liberal spirit of free research fostered by Burckhardt and his contemporaries. There was then no contradiction between Edelstein's endeavor to preserve what he considered good and valuable and his disregard, as a scholar, for biases and unfounded assumptions.

In the field of ancient medicine and science, to which I have to limit myself, this disregard led to startling insights. Departing from the customary interpretation of Asclepius as an aboriginal deity, Edelstein showed his appearance as a patron god of physicians prior to his metamorphosis into the healing god and, finally into the pagan savior. Having freed himself from the picture of the Hippocratic physician as a rudimentary scientist, Edelstein arrived at a new view of the relationship between medicine and philosophy. The scientific impetus came from the philosophers; here the physician was the recipient, though he may have taken over certain domains of research, such as the study of human anatomy and physiology. The scientific need for such study was a consequence of Aristotelian research; its realization was accomplished by physicians like Herophilus and Erasistratus and their heirs. But research in antiquity always remained the activity of a few, supported by the munificence of a few. It did not become a public concern, nor was it a common philosophic postulate. The academy doubted the possibility of comprehending nature, and this doubt was reflected in the school of the Empiricists, who rejected the quest for the "hidden" causes. The skepticism of Pyrrho, which extended over all causality, tinged the school of the medical Methodists. They deflected the atomism of Asclepiades and the "communities" of Themison into a doctrine of three classes of disease that easily revealed themselves by the symptoms the patient presented. This thesis of Edelstein's demanded an audacious readjustment of commonly accepted chronological data regarding the founders of the Methodist school.

Yet if ancient medicine was bound to ancient philosophy, Edelstein also perceived its role as a model for the philosopher. The philosopher, who insisted on being man's intellectual and moral guide, could and did point to man's willingness to entrust his body to the physician. Philosophy ought to be the medicine of the soul, just as medicine was the philosophy of the body.

Not a historian of medicine in the usual sense, Edelstein aimed at a history of man's scientific endeavors, *Wissenschaftsgeschichte,* to use the convenient German term. The above allusions to some of his contributions are meant to indicate his insistence on seeing a historical phenomenon in its own right, not just as an illustration of some general laws derived from sociology or anthropology. Nowhere does this become clearer than in his work on medical ethics. Because the Hippocratic Oath is still a stone in the foundation of medical ethics, we like to read timelessness into its rules. Edelstein interpreted the oath as a historical document. He asked himself whose ethics and practices were reflected and when this document was likely to have been written. In answer, he attributed the oath to a group of physicians oriented toward the Pythagorean way of life. He pleaded for its relatively late composition and its even later entry into the common consciousness of the ancient world. The Hippocratic Oath did not represent Greek medical ethics as such; it represented a phase in the history of ancient medical ethics, which Edelstein traced from the Hippocratic Corpus to late antiquity, in close relationship to the general history of ancient ethics.

The ethics of the medical profession had a history revealing aspects often strange to us. Yet behind the differences of conditions, rules, and etiquette, Edelstein also found something else: men striving for moral standards and, in so doing, imposing upon themselves restrictions as well as obligations. For years, Edelstein gave a lecture on the Hippocratic Oath to our medical students, a lecture that was received enthusiastically. Of medical science Edelstein knew little, I am afraid. But having seen much sickness in his family and himself having often been in the hands of doctors, he took a strong interest in the relationship of physician and patient. He had understanding for the medical man who felt scientific research to be his true vocation. But he held little sympathy for those who turned to science because of their contempt for bedside medicine or because of their unwillingness to take upon themselves the responsibilities and hardships of medical practice. Healing, he believed, was a noble profession.

As a lecturer, Edelstein had few equals. I made his acquaintance thirty-five years ago, when, upon Dr. Sigerist's invitation, he gave a lecture on Hippocrates at the Leipzig Institute of the History of Medicine. He spoke for about two hours without notes or manuscript. The audience, which included mature scholars of considerable reputation, was spellbound, in spite of the fact that the opinions he expressed must have run counter to many a man's convictions.

Edelstein would drive himself mercilessly to be clear in what he had to say, and I cannot remember any lecture he gave that was not an intellectual enrichment to the hearers, though his soft voice could make understanding difficult. He possessed a quiet humor, and his radiant yet gentle smile convinced even those of his hearers for whom the line of thought may have been too subtle.

There are many good lecturers. What singled out Edelstein as a great teacher was his fundamental seriousness—the fact that he offered his pupils bread, not stones. He would teach only what he believed in, what he could defend by deeds as well as by words. Edelstein knew that students have to be trained in the methods of their discipline and that the teacher has to pass on skills that cannot easily be learned from books alone. As a philologist he taught the art of interpretation in which he excelled. But teaching to him was more than training; the true teacher must stand to his pupils in a relationship of trust, where everything can be argued, but where the teacher's integrity must never be in doubt. This was one of the reasons for Edelstein's refusal to sign the special oath demanded by the regents of the University of California from its professors. How could he expect students to believe that he would always say what he thought to be true if he had shown his willingness to preserve his position at the price of having his beliefs censored? The trust between him and his students was necessary because Edelstein taught the humanities not as a number of disciplines traditionally passed on by this name, but as revelations of man's moral nature. Historical conditions changed, and opinions changed with them, even regarding obligations and privileges, as he had demonstrated in the history of Greek medical ethics. Yet behind the changing codes and behind the "otherness" of ancient civilization, there was the ever remaining demand on the individual to make decisions that were good or bad. Edelstein perceived man's moral nature in the ability of reason to say "no" to the drives that ruled the animal. To the

best of my knowledge, Edelstein never accepted the theory of man's animal descent. Far from being a rationalist in the sense of denying the claims of human feelings, he yet maintained a faith in man's essential reasonableness. He was fond of engaging persons of all walks of life in conversation, and I remember his telling how he had made somebody in California see the merits of his side in the oath controversy, in spite of the widespread prejudice against it. He took such cases as proof that it was not necessary to declare impossible what reason insisted was right. I was not sure whether he realized that, in such cases, reason was often helped by having Edelstein's personality and faith for its speaker. Or perhaps he did realize it and would not have it otherwise. For faith, to him, was an essential partner of reason. Only to the man without faith, who easily yielded to stronger forces, were historical events determined in advance. When arguing a case he believed just, Edelstein behaved much as he did when interpreting Plato, for whom he had a reverent attachment. Something of the Socratic suavity, irony, and incorruptibility seemed to come to life again and once more to prove convincing to pupils and strangers alike. And so he had faith in the existence of a "truth" that the scholar must seek.

"The relativity and historicity of so many values must not deceive us into believing that no standards are generally accepted. The light of truth, broken as it is into the spectrum of individual existence, still is the reflection of the light of a truth absolute, not relative, everlasting and not waning. Through discussion, through argument, through ever renewed consideration of problems, we do find the better in the course of time— as much of the truth as mortal eyes are able to discover."[2] These words, taken from Edelstein's Phi Beta Kappa address at Columbia University, seem to me to reflect his philosophy as a scholar.

In Ludwig Edelstein we have lost a scholar who contributed much to our understanding of man's search for knowledge, a scholar who taught us that the small detail and the last aim are all of one piece, a scholar who was a moral force because he told us not to compromise with what we think wrong and because he tried to live what to him seemed right. To all of us, as well as to those after us, he has left his work, and to his friends he has left the memory of a man gentle when they suffered, helpful when they needed help, and generously loyal to all he deemed good in them.

NOTES

This tribute was presented at the Johns Hopkins Institute of the History of Medicine, November 23, 1965.

1. Ludwig Edelstein, "Philosophy—the pilot of life," *Teachers College Rec* 65 (January 1964): 302.

2. Ibid., 306.

BIBLIOGRAPHY

This bibliography lists Dr. Edelstein's publications that appeared during his lifetime. In view of the great interest among active scholars for Dr. Edelstein's work, it did not seem advisable to defer this compilation until all possible posthumous publications were available. . . . The help of Dr. Harry Frankfurt of the Rockefeller University and of Dr. Harold Cherniss of the Institute for Advanced Study is gratefully acknowledged.

The bibliography was prepared with the assistance of Mrs. Janet B. Koudelka.

Books

Περὶ ἀέρων *und die Sammlung der hippokratischen Schriften* (Berlin: Weidmannsche Buchhandlung, 1931) (Problemata, Heft 4).

The Hippocratic Oath: Text, Translation, and Interpretation (Baltimore: Johns Hopkins Press, 1943) (Supplements to the *Bulletin of the History of Medicine*, no. 1).

(With Emma J. Edelstein). *Asclepius: A Collection and Interpretation of the Testimonies*, 2 vols. (Baltimore: Johns Hopkins Press, 1945) (Publications of the Institute of the History of Medicine, Second Series: Texts and Documents, vol. 2).

Henry R. Zimmer. *Hindu Medicine*, edited with a foreword and preface by Ludwig Edelstein (Baltimore: Johns Hopkins Press, 1948) (Publications of the Institute of the History of Medicine, Third Series: The Hideyo Noguchi Lectures, vol. 6).

Erich Frank. *Wissen, Wollen, Glauben; gesammelte Aufsätze zur Philosophie-geschichte und Existentialphilosophie* (Zürich: Artemis-Verlag, 1955) (Erich Frank, *Knowledge, Will, and Belief: Collected Essays*, ed. Ludwig Edelstein; distributed in the United States by Henry Regnery Co., Chicago).

Plato's Seventh Letter (Leiden: Brill, 1966) (Philosophia Antiqua) [in press].

Articles

"Antike Diätetik," *Die Antike* 7 (1931): 255–70.

"Die Geschichte der Sektion in der Antike," *Quellen und Studien zur Geschichte der Naturwissenschaften und der Medizin* 3, no. 2 (1932): 50–106.

"Empirie und Skepsis in der Lehre der griechischen Empirikerschule," *Quellen und Studien zur Geschichte der Naturwissenschaften und der Medizin* 3, no. 4 (1933): 45–53.

"Cicero: 'De natura deorum,' II," *Studi italiani di filologia classica* 11, no. 2 (1934, n.s.): 131–83.

"The development of Greek anatomy," *Bull Hist Med* 3 (1935): 235–48.

"Methodiker: Pauly-Wissowa," *Realenzyklopädie der klassischen Altertumswissenschaft*, Supplementband 6 (1935): cols. 358–73.

"Hippokrates von Kos: Pauly-Wissowa," *Realenzyklopädie der klassischen Altertumswissenschaft*, Supplementband 6 (1935): cols. 1290–1345.

"The philosophical system of Posidonius," *Am J Philology* 57 (1936): 286–325.

"Winckelmann and the translation of Dioscurides," *Bull Hist Med* 4 (1936): 261–63.

"Greek medicine in its relation to religion and magic," *Bull Hist Med* 5 (1937): 201–46.

"The genuine works of Hippocrates," *Bull Hist Med* 7 (1939): 236–48.

"Platonism or Aristotelianism? A contribution to the history of medicine and science," *Bull Hist Med* 8 (1940): 757–69.

"Primum Graius Homo (Lucretius 1.66)," *Trans Am Philological A* 71 (1940): 78–90.

"Horace, Odes II, 7, 9–10," *Am J Philology* 62 (1941): 441–51.

"Andreas Vesalius, the humanist," *Bull Hist Med* 14 (1943): 547–61.

"Aristotle and the concept of evolution," *Classical Weekly* 37 (1944): 148–50.

"Sydenham and Cervantes," *Bull Hist Med* Supplement 3 (1944): 55–61.

"The role of Eryximachus in Plato's Symposium," *Trans Am Philological A* 76 (1945): 85–103.

"William Osler's philosophy," *Bull Hist Med* 20 (1946): 270–93.

"Medical historiography in 1847," *Bull Hist Med* 21 (1947): 495–511.

[Articles on Greek medicine], in *Oxford Classical Dictionary* (Oxford: Clarendon Press, 1949); s.v. Asclepiades, Athenaeus, Dentistry, Dietetics, Diocles of Carystus, Dioscorides Pedanius, Erasistratus, Galen, Hippiatrici, Hippocrates, Herophilus, Ointments, Ophthalmology, Philinus of Cos, Philistion of Locri, Physiognomonici, Praxagoras of Cos, Soranus of Ephesus, Rufus of Ephesus, Themison of Laodicea, Thessalus of Tralles.

"The function of the myth in Plato's philosophy," *J Hist Ideas* 10 (1949): 463–81.

"Wielands 'Abderiten' und der deutsche Humanismus," *Univ Calif Publ Modern Philology* 26, no. 5 (1950): 441–72.

"The interpretation of Erich Frank by Santillana and Pitts," *Isis* 43 (1952): 119–21.

"Recent trends in the interpretation of ancient science," *J Hist Ideas* 13 (1952): 573–604.

"The relation of ancient philosophy to medicine," *Bull Hist Med* 26 (1952): 299–316.

"The golden chain of Homer," in *Studies in Intellectual History* (Baltimore: Johns Hopkins Press, 1953), 48–66.

"Erich Frank's work: an appreciation," in *Knowledge, Will, and Belief: Collected Essays*, ed. Ludwig Edelstein (Zurich: Artemis-Verlag, 1955) (distributed in U.S. by Henry Regnery, Chicago), 407–65.

"The professional ethics of the Greek physician," *Bull Hist Med* 30 (1956): 391–419.

Letters to the editor of the *Baltimore Sun*, Feb. 5, 1959, p. 16 ("Darwin and Darwinism") and Feb. 22, 1959, p. 18 ("Kant and Darwin").

"Some reflections on 'The two cultures and the scientific revolution,'" *Rockefeller Inst Q* 5 (1961): 1–4.

"Platonic anonymity: for George Boas, August 28, 1961," *Am J Philology* 83 (1962): 1–22.

"Randall on Aristotle, a review," *J Philosophy* 59 (1962): 151–66.

"In memory of A. O. Lovejoy," *J Hist Ideas* 24 (1963): 451–56.

"Motives and incentives for science in antiquity," in *Scientific Change: Historical Studies in the Intellectual, Social, and Technical Conditions for Scientific Discovery and Technical Invention, from Antiquity to the Present*, ed. A. C. Crombie (New York: Basic Books, 1963), 15–41.

"Philosophy—the pilot of life (Phi Beta Kappa address at Columbia University)," *Teachers College Record* 65 (1964): 299–310.

"The Greco-Roman concept of scientific progress," in *ITHACA, Actes du Xme Congrès International d'Histoire des Sciences 26 VIII–2 IX 1962* (Paris: Hermann, 1965), 47–59.

Reviews

Karl Deichgräber, *Die griechische Empirikerschule*, Sammlung der Fragmente und Darstellung der Lehre (Berlin, 1930), *Deutsche Literaturzeitung* 24 (1931): cols. 1113–20.

Margit Gutman, *Die Nebensätze in ausgewählten Schriften des hippokratischen Corpus und ihre Bedeutung für die Verfasserfrage*, diss. Munich, 1929, *Gnomon* 7 (1931): 328–30.

E. M. Butler, *The Tyranny of Greece over Germany: A Study of the Influence Exercised by Greek Art and Poetry over the Great German Writers of the Eighteenth, Nineteenth, and Twentieth Centuries* (Cambridge, 1935), *Modern Language Notes* 53 (1938): 53–56.

Pierre Boyancé, *Études sur le songe de Scipion* (Paris, 1936), *Am J Philology* 59 (1938): 360–64.

Lotte Labowsky, *Die Ethik des Panaitios* (Leipzig, 1934), *Am J Philology* 59 (1938): 99–102.

W. Stettner, *Die Seelenwanderung bei Griechen und Römern* (Stuttgart, 1934), *Am J Philology* 59 (1938): 503–5.

Wilhelm Dilthey, *Gesammelte Schriften*, vol. II, *Vom Aufgang des geschichtlichen Bewusstseins*, Jugendaufsätze und Erinnerungen (Leipzig: Herausgegeben von Erich Weniger, 1936), *Modern Language Notes* 53 (1938): 146–50.

Aristotle, *Parts of Animals*, English trans. A. L. Peck, and *Movement of Animals, Progression of Animals*, English trans. E. S. Forster, *Classical Weekly* 32 (1938–39): 78–79.

R. Strömberg, *Theophrastea*, Studien zur botanischen Begriffsbildung (Göteborg, 1937), *Isis* 31 (1939): 83–85.

W. Jaeger, *Diokles von Karystos: Die griechische Medizin und die Schule des Aristoteles* (Berlin, 1938), *Am J Philology* 61 (1940): 483–89.

M. Pohlenz, *Hippokrates und die Begründung der wissenchaftlichen Medizin* (Berlin, 1938), *Am J Philology* 61 (1940): 221–29.

Marie Delcourt, *Stérilités mystérieuses et naissances maléfiques dans l'antiquité classique* (Paris, 1938), *Classical Weekly* 34 (1941): 114–15.

William Arthur Heidel, *Hippocratic Medicine: Its Spirit and Method* (New York, 1941), *Classical Weekly* 35 (1941–42): 147–48.

L. A. Stella, *Importanza di Alcmeone nella storia del pensiero greco* (Rome, 1939), *Am J Philology* 63 (1942): 371–72.

K. Schubring, *Untersuchungen zur Überlieferungsgeschichte der hippokratischen Schrift De locis in homine* (Berlin, 1941), *Am J Philology* 65 (1944): 306–9.

Humphrey Trevelyan, *Goethe and the Greeks* (Cambridge, 1941), *Am J Philology* 66 (1945): 70–74.

F. H. Wagman, *Magic and Natural Science in German Baroque Literature: A Study in the Prose Forms of the Later Seventeenth Century* (New York, 1942), *Modern Language Notes* 60 (1945): 58–61.

William F. Petersen, *Hippocratic Wisdom: A Modern Appreciation of Ancient Scientific Achievement* (Springfield, Ill., 1946), *Bull Hist Med* 20 (1946): 475–82.

Lynn Thorndike, *The Herbal of Rufinus* (Chicago, 1945), *Modern Language Notes* 61 (1946): 355–56.

Galen on Medical Experience, 1st ed. of Arabic version with English trans. and notes by R. Walzer (London, 1944), *Philosophical Rev* 56 (1947): 215–20.

Modestus van Straaten, *Panétius: Sa vie, ses écrits et sa doctrine avec une édition des fragments* (Amsterdam, 1946), *Am J Philology* 71 (1950): 78–83.

Erich Auerbach, *Mimesis: Dargestellte Wirklichkeit in der abendländischen Literatur* (Bern, 1946), *Modern Language Notes* 65 (1950): 426–31.

Max Pohlenz, *Die Stoa: Geschichte einer geistigen Bewegung* (Göttingen, 1948), *Am J Philology* 72 (1952): 426–32.

Pierre-Maxime Schuhl, *Essai sur la formation de la pensée grecque: introduction historique à une étude de la philosophie platonicienne*, 2nd ed. (Paris, 1949), *Classical Weekly* 45 (1952): 184, 186.

Henri Grégoire, avec la collaboration de R. Goosens et de M. Mathieu, *Asklépios, Appollon Smintheus et Rudra: Études sur le dieu à la taupe et le dieu au rat dans la Grèce et dans l'Inde* (Bruxelles, 1950), *Gnomon* 26 (1954): 162–68.

Fritz Wehrli, *Die Schule des Aristoteles, Texte und Kommentar*, Heft 4, *Demetrios von Phaleron*; Heft 5, *Straton von Lampsakos*; Heft 6, *Lykon und Ariston von Keos*; Heft 7, *Herakleides Pontikos* (Basel, 1949–53), *Am J Philology* 76 (1955): 414–22.

Index

abortion: as ethical question, 15; forbidden by Declaration of Geneva, 29; in Hippocratic Oath, 24; Soranus on, 30–31, 35; suppository for, 27n. 4; Tertullian on, 35, 46n. 26

Ackerknecht, Erwin H.: on advanced age, 1; on medical ethics, 42

administration, medical care, 208

Agricola, Georg: *Bermannus*, 66; on mining, 65–66; on standards, 66

Agrippa of Nettesheim, Henricus Cornelius: *On Occult Medicine*, 68–69; magic related to religion of, 82n. 23

Alberti, Leon Battista, on education, 71

alchemy, 188–89

alcoholism, as disease, 49

ancient medicine: "otherness" of, 2; study of, 1–6

animalcular theory of disease, 216

animals, man's relationship to, 92–94

antibiotics, definition of, 165, 176, 177n. 1

antiquarianism, one-sided, 244

apprenticeship, art learned by, 235

Apuleius, *The Golden Ass*, 31–32

Arbuthnot, John, *Essay Concerning the Nature of Aliments*, 191

Aristotle, 70–71

Arnold, headmaster of Rugby, 71–72

arsphenamine for syphilis, 171

artist, scientific knowledge of, 67

Asclepiades, on drug therapy, 154

Asclepiads, Hippocratic Oath sworn by, 27n. 8

Asclepius, as patron god of physicians, 21, 22, 255

astronomy, of Copernicus, 77–79

Autenrieth, J. H. F. v., biography of, 199

Baltimore, climate of, 201

Beethoven, deafness of, 54

Bernard, Claude: on arts vs. sciences, 222; experimental pharmacology of, 159, 175; *Introduction to the Study of Experimental Medicine*, 224; on scientific truth, 139, 140; on statistics, 212

bibliography of Ludwig Edelstein, 259–63

Bichat, 107; as sensualist, 97

Bigelow, "On Self-Limited Diseases," 155

biography: and historiography, 6–10; by medical historians, 8

bladder stones, lithotomy for, 23

Bodin, Jean, *Method for the Easy Understanding of History*, 133

books, historical value of, 221

Bouillaud, J., as Gall's supporter, 91

brandy, in Bamberg Hospital, 173

Brecht, on Galileo, 144

Broussais, F.-J.-V., 109; attitude toward Gall, 101–2; *De l'irritation et de la folie*, 99–100; pathology reformed by, 104; as phrenologist, 101; on physiologists vs. spiritualists, 100

Brown, John: Rush's yellow fever treatment and, 198–99; theory of disease of, 196–97